Computing for Scientists

THE MANCHESTER PHYSICS SERIE

General Editors: D. J. SANDIFORD; F. MANDL; A. C. PHILLIPS
Department of Physics and Astronomy, University of Manchester

PROPERTIES OF MATTER	B. H. FLOWERS and E. MENDOZA
OPTICS Second Edition	F. G. SMITH and J. H. THOMSON
STATISTICAL PHYSICS Second Edition	F. MANDL
ELECTROMAGNETISM Second Edition	I. S. GRANT and W. R. PHILLIPS
STATISTICS	R. J. BARLOW
SOLID STATE PHYSICS Second Edition	J. R. HOOK and H. E. HALL
QUANTUM MECHANICS	F. MANDL
PARTICLE PHYSICS Second Edition	B. R. MARTIN and G. SHAW
THE PHYSICS OF STARS	A. C. PHILLIPS
COMPUTING FOR SCIENTISTS	R. J. BARLOW and A. R. BARNETT

Computing for Scientists
Principles of Programming with Fortran 90 and C++

R. J. Barlow and A. R. Barnett
Schuster Laboratory, University of Manchester

JOHN WILEY & SONS
Chichester · New York · Weinheim · Brisbane · Singapore · Toronto

Other Wiley Editorial Offices

John Wiley & Sons, Inc., 605 Third Avenue,
New York, NY 10158-0012, USA

WILEY-VCH Verlag GmbH, Pappelallee 3,
D-69469 Weinheim, Germany

Jacaranda Wiley Ltd, 33 Park Road, Milton,
Queensland 4064, Australia

John Wiley & Sons (Asia) Pte Ltd, 2 Clementi Loop #02-01,
Jin Xing Distripark, Singapore 129809

John Wiley & Sons (Canada) Ltd, 22 Worcester Road,
Rexdale, Ontario M9W 1L1, Canada

Antony Rowe Ltd,

British Library Cataloguing in Publication Data

A catalogue record for this book is available from the British Library

ISBN   0 471 95114 5

Produced from camera-ready copy supplied by the authors

For the three generations   Stan, Fiona;   Pat;   Alex and Jessica – A.R.B.

For Edward – R.J.B.

Contents

6   CHARACTERS AND STRINGS

Editors' preface to the Manchester Physics Series

The Manchester Physics Series is a series of textbooks at first degree level. It grew out of our experience at the Department of Physics and Astronomy at Manchester University, widely shared elsewhere, that many textbooks contain much more material than can be accommodated in a typical undergraduate course; and that this material is only rarely so arranged as to allow the definition of a shorter self-contained course. In planning these books we have had two objectives. One was to produce short books: so that lecturers should find them attractive for undergraduate courses; so that students should not be frightened off by their encyclopaedic size or their price. To achieve this, we have been very selective in the choice of topics, with the emphasis on the basic physics together with some instructive, stimulating and useful applications. Our second objective was to produce books which allow courses of different lengths and difficulty to be selected, with emphasis on different applications. To achieve such flexibility we have encouraged authors to use flow diagrams showing the logical connections between different chapters and to put some topics in starred sections. These cover more advanced and alternative material which is not required for the understanding of later parts of each volume.

Although these books were conceived as a series, each of them is self-contained and can be used independently of the others. Several of them are suitable for wider use in other sciences.

The Manchester Physics Series has been very successful with total sales of more than a quarter of a million copies. We are extremely grateful to the many students and colleagues, at Manchester and elsewhere, for helpful criticisms and stimulating comments. Our particular thanks go to the authors for all the work they have done, for the many new ideas they have contributed, and for discussing patiently, and often accepting, the suggestions of the editors.

Finally, we would like to thank our publishers, John Wiley & Sons, for their enthusiastic and continued commitment to the Manchester Physics Series.

D J SANDIFORD

F MANDL

January, 1998   A C PHILLIPS

0

Preamble

Of making many books there is no end
— Ecclesiastes 12:12

In which the authors justify the production of yet another publication on computing, and explain what the book tries to do and the audience for whom it is appropriate.

The need for a new book on programming might seem questionable in view of the enormous number of volumes now being published on the subject. Nevertheless we think we have something to offer, meeting a need which is not met by the 9999 other titles available.

These books – many of them excellent – tell the reader *how* to program in a particular language. They are like books of recipes: if you perform certain actions (mixing flour and sugar, or declaring variables to be high precision) you will get certain desirable results (a chocolate sponge cake, or an accurate calculation.) We aim to go deeper than that. To explain something of the ways that programs are actually implemented and what happens when they run. To change metaphor, this is like teaching a learner driver not merely about the use of the accelerator, clutch and gear-lever, but to open the bonnet of the car* and show them the engine at work.

---

* Or, if you prefer, the hood of the automobile.

Such deeper appreciation of what's happening is vital if a programmer is to get beyond a surface ability to a true understanding: to appreciate *why* programs work (or work better) when written in particular ways. The underlying theme of the book is the development of the *thoughtful programmer* who writes better programs, and gets more fulfilment out of doing so. We assume that a scientific programmer wants to develop his or her programming knowledge and skill in this way, albeit without attaining the full expertise necessary for a professional software writer.

The problem with the 'recipe book' method of learning is that the programmer is very reluctant to try anything new, beyond the bounds of what is familiar. They stick to a pattern which they know to work because it is familiar and safe – that's why, to take an example, some programs are full of baroque **if–then–else** constructions when a **switch** or **CASE** would be more appropriate; others use **switch** or **CASE** for two-way choices that would be better served by **if–then–else**. It is particularly deadly when programmers are adopting other people's programs rather than writing their own from scratch – calls and constructions are left in even though they serve no useful purpose, as part of a superstitious ritual. We try and show that in programming – unlike physics – there is no unique 'correct solution' to a problem: a thoughtful programmer will not just choose a technique to write a program to solve a problem, but consider several techniques and choose an appropriate one for this instance.

We have been writing scientific programs for many years, in several languages and on diverse systems. This provides us with a suitably long perspective to discuss today's computing. It helps us help our students – we can spot a wrongly typed argument in a **Fortran** subroutine call, or a forgotten **break** in a **switch** construction because we ourselves have made such mistakes so often and so painfully. We hope that some of this accumulated experience comes through in the following pages. We also found that – in early drafts – it produced an undue amount of anecdote; this we have done our best to suppress.

We stress throughout the importance of programming style to produce clear and readable programs. This is born out of our years of painful experience tracing through programs (our own and other people's) to try and discover just what they do – and why they're going wrong. There is no single prescription for good style, though various sets of rules can be found in the literature; we do not believe that any of these 'systems so perfect that no one* will need to be good' relieves the programmer of all other responsibility. In our **C++** code we have adopted the system of indentation whereby all curly brackets { braces } appear on a line of their own. Other systems are possible and we would not claim that this one is 'best' – what matters is that the programmer writes the source code in such a way that the block structure is clear to the reader.

The need to cram programs into two columns of text did constrain our programming style. Not only are the indentations smaller than we would use in practice, we also found we had to use variables with short names to get the code to fit into the format

---

* i.e. no programmer.

– in contradiction to our own advice to 'resist the temptation to call all integers **i** and all floating-point variables **x**'. We hope the critical reader will understand the circumstances, and not follow this brief style when they write their own programs with the luxury of a full 80-column screen.

We use two languages so as to bring out ideas rather than instances. The perspective of bilingualism lets the reader separate the basic concept (for example, iteration) from the way it is implemented in the language (**DO** loops and **for** loops). This helps them understand what's really happening, and also to appreciate more easily any similar constructions in other languages and systems. We do not adopt the particular viewpoint of either language – this is not a book about **Fortran** or about **C++** , but a book about computing, written using the languages of **Fortran** and **C++** .

A book such as this may be read from beginning to end, as part of a programming course or for self-study, but very often a reader picks it up wanting to look at a particular topic. (Thus it is used for both sequential and for random access.) So we have avoided the technique of referring backwards ('... as we saw earlier') and of building up concepts and ideas by means of long-running examples. We find these quite infuriating when we try and look things up in other people's books, and resolved to spare our own readers.

The fact that the reader of section n cannot be assumed to have read and remembered sections $1 \ldots (n-1)$ leads inevitably to a certain amount of repetition of material. We have done our best to keep this to the necessary minimum and we trust readers will tolerate it when they come across it. In particular the various data types are discussed briefly in Chapter 3 and are then revisited in later chapters. This structure provides a minimum necessary to get started at the early stages, saving the more complete details till later. The price paid is a lot of forward and backward referencing and a certain amount of overlap.

It is impossible to cover either language completely in a single short book, and so we have not attempted to do so. We have used this freedom to leave out many features, both small and large, using the grounds of obscurity or of poor style. There is no mention of **Fortran COMMON** blocks – despite their importance in previous versions of **Fortran** – for their potential for devastating confusion between subprograms makes them far inferior to the system of global declarations now available in **Fortran 90**. Likewise the **C** preprocessor gets only the most casual mention, as the **C++ const** is supposed to replace the **#define** command – though we suspect this will take a long time as the older way of doing things is well entrenched (and faster). **union**s are not mentioned at all, and neither is the **EQUIVALENCE** statement; their main use is in converting data between types, which is not something to be encouraged.

So this book is not a substitute for the appropriate manual for the reader's particular language and particular system. There they can find (if they look in the right place, which isn't always easy given the appalling indexing technique of some manual writers) details of the obscure or unhygenic features we neglect, the details of the idiosyncratic additions the vendor has made to the language standard, and an account of what commands are used to compile and link and run programs – any attempt on our part to cover these (we did consider it) would be inevitably incomplete and obsolescent. We do hope, however, that use of our book will make the thick manual easier to understand, and easier to navigate through.

We would like to express our thanks to our colleagues who have read and commented on drafts of this work, especially James Youngman and Peter Mitchell. Their knowledgeable criticisms have added very significantly to the end result, and we greatly appreciate the time and effort they have given. Willing and valuable help from Sze Tan greatly helped at a critical stage.

It should go without saying that any remaining errors and omissions are our responsibility; we encourage readers who find them to communicate with us using the email addresses **Roger.Barlow@man.ac.uk** and **Ross.Barnett@man.ac.uk**. The web page **http://www.hep.man.ac.uk/~roger/programming.html** contains notes of any such errata, and also machine-readable versions of complete code in the book, although we do not encourage too much reliance on our particular versions of the principles we expound. (i.e. don't just copy our stuff: read the book and write your own!)

We are grateful for permission to use the quotation from Robert Frost from *The Poetry of Robert Frost*, edited by Edward Connery Latham, the Estate of Robert Frost, published by Jonathan Cape.

Ross Barnett
Roger Barlow
November 1997

Albert's uncle says I ought to have put this in the preface, but I never read prefaces, and it is not much good writing things just for people to skip. I wonder other authors have never thought of this.
– E. Nesbit: The Story of the Treasure Seekers

1

Basic Concepts

An introductory view of the different types of program, the basic features of what goes on in the hardware, and how a program is transformed from written text to executable instructions. Guidance for writing readable and useful programs is given, and the two languages chosen for the book are discussed. The programs that provide the computing environment are briefly explained.

Computers are everywhere. In homes, in schools, universities, and business offices, they can be found in all shapes and sizes, from small personal computers that slip into a briefcase to large supercomputers that fill a room.

You have probably – almost certainly – used computers in some way. Perhaps for some simple numerical work of a predefined pattern, perhaps for word-processing, perhaps only for playing games. Indeed, they can be used very effectively at this level. If you have some common problem, such as plotting a straight line through a set of data points, or solving a set of linear simultaneous equations, or predicting your firm's sales over the next 5 years, then somebody has probably already created a piece of applications software that is just what you require. All you have to do is run it and follow their instructions. Many computer users are happy to stop there. Their needs are all provided by existing software packages, and they never write, or need to write, any software of their own.

If you are one of these people, and want to go no further, you have bought the wrong book.

1.1 HARDWARE, SOFTWARE AND PROGRAMS

The physical machine defines the *hardware*. In the certainty of being dated within 6 months let us specify a state-of-the-art PC at the start of 1998. Such a machine has a processor with a clock speed of 233 MHz, 32 Mbytes of RAM, a 4-Mbyte graphics card, a 4-Gbyte hard disc, and probably a 12-speed CD-ROM, with other multimedia devices; a modem to connect you to email and the World Wide Web is common.*

The hardware will remain inert and useless without *software programs* which run on it. One such program is the *operating system* (§1.12) which controls the running of other programs and everything else that goes on in the machine. There are stand-alone *packages* such as spreadsheets, word processors, text-editors, CAD programs, and physics-teaching programs, which isolate you from their interior and which provide an environment for you. There are *languages* – **Fortran**, **C++**, Pascal, Assembler, Ada, Basic ... There are *computing environments* for mathematics and science, such as Mathematica^{TM}, MATLAB^{TM}, MathCAD^{TM}, and Reduce^{TM} which contain their own language conventions. And finally there are *your* programs which are written in one of the languages and over which you have absolute control.

With so much software already written and available, you might think that there was no point in writing your own: that programming should be left to the skilled professionals, with no place for the 'do-it-yourself' amateur. But there are reasons why writing your own software can be not just an indulgence but a real asset:

- some problems may not be handled by your available software. Data may require analysis in a particular way that doesn't fit easily into the package.

- you may not completely trust the way a package analyses your data, and want to verify the results or try a different graphical slant.

- knowing about software principles will help you understand the packages and get the best out of them.

- you may want to become a professional (or semi-professional) software writer.

- as a curious scientist you want to find out how computers work, rather than just accept their results.

The purpose of this book is to enable you to understand how a computer works, and how to write your own software to solve your scientific problems. Scientific programming is concerned with mathematical computations: with setting up and solving models of the physical world, and also with the analysis of data: finding

---

* Just over 10 years ago (mid-1980s) we were doing useful computing with something the size of a BBC Micro, whose entire memory was 64 kbytes, which is 5000 times smaller than those of today! We doubt that we do 5000 times more work today or even work 5000 times more quickly ...

solving models of the physical world, and also with the analysis of data: finding meaningful patterns in the complicated data taken by modern experiments. It is thus very much concerned with numerical calculations and how to perform them efficiently and correctly, and also with the presentation of the results in a sensible way, often in graphical form.

The chapters of the book contain various programs, or fragments of programs. These have been chosen to illustrate particular programming concepts, but we hope that they will be of some relevance to you and the problems that you want to tackle now, or will want to in the future.

1.1.1   Two languages

We have chosen two languages to work with: **Fortran** and **C++**. These are good examples of modern computer languages, and are widely available. But don't think that this is a book about **Fortran**, or about **C++** : it is about computing, using **Fortran** and **C++** to illustrate the concepts involved in writing software.

The present **Fortran** language is called **Fortran 90**. The language design was begun in 1980 and was expected to culminate in that decade, hence the draft was called FORTRAN 8X. In fact the **Fortran** ANSI* standard, ANSI(X3-198 1992), appeared finally in 1992. It is described in Adams et al. (1992).

The **C++** language standard has fared no better. **C++** was created in 1986 and is closely based[†] on the **C** language. The **C++** ANSI standard became available in December 1997, although a draft was published in 1994. The **C** language itself was invented in 1970 and the ANSI standard duly followed – in 1989! Its bible, as a language definition, is the book by Kernighan and Ritchie (1988).

Adhering to the language standard, rather than using convenient features peculiar to your own system, is important in serious programming: writing portable code that will perform correctly on several platforms is a worthwhile goal and it requires dedication and considerable knowledge.

We assume that you have access to a computer system which supports one (or, even better, both) of these two languages, **Fortran** and **C++**. These days compilers are available for the PC end of the market as well as on Unix systems (generally workstations). That may just mean that a compiler and linker are provided somewhere on your system; more probably it will provide an integrated environment with a language-specific editor, from which the compiler and linker can be invoked, and your programs can be run. Exactly how this is achieved will depend on the particular hardware and software at your disposal: we will do our best to guide you through the process, but to check on specifics you will need the manual for your particular system within easy reach.

---

* ANSI is the American National Standards Institute, and is generally accepted as providing the definitive standard for computer languages.
† Where C and C++ cover the same ground they are almost (but not quite) identical, but C++ is greatly extended in scope. One discussion report was called *As close to C as possible – but no closer.*

1.2 INSIDE THE BOX

If you take the lid off your computer – having remembered to disconnect the power supply first – then what you see inside will not at first give any immediate clue as to how it works. You will find lots of integrated circuits (*chips*), joined by a great many connections made via printed-circuit boards.

Some of these chips, probably a large block all together, comprise the memory (or 'RAM' – *Random Access Memory*) of the computer. If you could take them apart and examine them under a microscope, you would discover that they consist of hundreds of thousands of individual transistors, made by etching and doping a slice of silicon crystal. In operation, these transistors are each either in the OFF (non-conducting) or ON (conducting) state. Each thus contains one piece of information (called a *bit*), and it is this data that provides all the information that controls the actions of the computer.

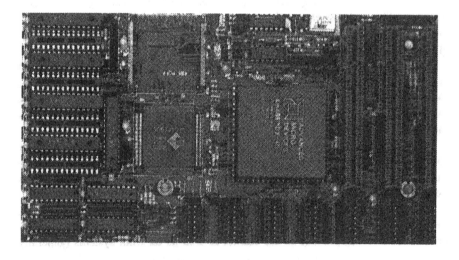

Fig. 1.1   Part of a typical board

An unstructured collection of bits is not enough to form any useful information; the way the bits are arranged is equally vital – just as information in English comes not only in the letters it contains, but in the way they are arranged into words and sentences. (If you cut up this page into little pieces, with one letter on each piece, and mix them all together, then the result wouldn't tell you much. Even this instruction would be lost!) This structured arrangement happens in two stages: the construction of bytes and words (§1.2.1), followed by their storage and retrieval at specified addresses in memory (§1.2.2).

1.2.1 Bits, bytes and words

One bit contains the minimum information possible, so as a first stage of organisation the data bits are put together in groups of 8 called *bytes* and in groups of 32 (usually: some hardware systems have a different number) called *words*.

The bits within a byte are usually numbered from 0 to 7 so that they can be distinguished from each other. The contents can then be given precisely by specifying whether the 8 micro transistors involved are in their ON or OFF state. A typical set of byte contents is shown in Table 1.1. A more compact way of giving the same information is to write 0 for OFF and 1 for ON and get the 8 numbers in the bottom row.*

Table 1.1 Bit pattern for a computer byte

Bit	0	1	2	3	4	5	6	7
State	OFF	ON	OFF	ON	OFF	OFF	ON	ON
Binary number	0	1	0	1	0	0	1	1
Place value	128	64	32	16	8	4	2	1

- **Binary numbers**
  This bit pattern can then be interpreted as the binary number 01010011_2: a number expressed according to the base 2, where the place values of the digits are not the powers of 10: 1, 10, 100, 1000... as in normal base-10 decimal arithmetic, but powers of 2: 1, 2, 4, 8, 16... Just as there are no digits larger than $9(= 10 - 1)$ in our normal arithmetic, in binary there are no digits larger than $1(= 2 - 1)$.

- **Decimal numbers**
  The number can be written still more compactly in the usual base-10 decimal notation. (It is $64 + 0 + 16 + 0 + 0 + 2 + 1 = 83_{10}$.) But going from one to another is a rather messy process. To get from binary to decimal you have to do the sum above and decimal to binary is worse. (Divide repeatedly by 2, writing down the remainder, till nothing is left.)

With 32-bit words a compact notation is even more essential, but decimal to binary conversions are even more unfriendly with large numbers: it's hard to answer questions like 'Is bit 7 on or off in the number 862343_{10}?' A suitable compromise is the use of octal (base 8) and/or hexadecimal (base 16) notation.

---

* Or, in some systems, the other way round! It doesn't matter which provided you are consistent.

- **Octal numbers**

  In *octal* the place values are the powers of 8: 1, 8, 64, 512, 4096... The digits that can occur are 0 to 7. Each octal digit requires 3 binary digits, so 8 bits fit with room to spare into 3 octal digits, ranging from 000_8 to 377_8. Even 32 bits – the full word – can be expressed as 11 octal digits. Translation from binary to octal is easy: you just group the binary digits together in threes, starting from the right, and each group corresponds to the appropriate octal digit.

Table 1.2   Octal form of the binary bit pattern 10110010000110101100110110011101

Grouped binary	10	110	010	001	101	011	100	110	110	011	101
Octal number	2	6	2	1	5	3	4	6	6	3	5

- **Hexadecimal numbers**

  Hexadecimal ('hex') uses powers of 16, and each digit corresponds to 4 bits. So 8 digits are enough to represent a word. New symbols are needed as hex digits representing the numbers 10 to 15, and the letters A to F are used for this.

Table 1.3   Hexadecimal form of the bit pattern 10110010000110101100110110011101

Grouped binary	1011	0010	0011	0101	1100	1101	1001	1101
Hexadecimal number	B	2	3	5	C	D	9	D

Hence the number in Tables 1.2 and 1.3 can be written in binary, octal and hex $10110010000110101100110110011101_2 = 26215346635_8 = B235CD9D_{16}$.

The bit patterns for octal and hex digits soon become familiar. A calculator that converts between decimal and octal and hex is an excellent investment. You may not need to use it very often, but when you do, it's worth it. (But be careful! The 'B' symbol can easily be mistaken for a '6' in some models.)

Problem 1.1   *In some systems the number 3735943886 is used to fill unused areas of memory. 'Hexplain' why that number should be chosen.*

Despite this numerical emphasis, please note that it is wrong to say that 'the computer stores binary numbers'. The computer stores bit patterns which can be used to represent binary numbers. They can also be used to represent other things, as we shall see: floating-point numbers, alphabetic characters, machine-code instructions... The bit pattern in a word can always be expressed as a binary number: whether the meaning of the pattern is that number or something else depends on the intentions of the person or program that put it there. Many of the most dramatic programming errors are caused by a data pattern being stored according to one interpretation and then read out according to a different and inconsistent interpretation.

1.2.2   Words and memory

The second stage of hardware organisation is that each of these words has an address: a unique number which identifies it. Thus a 1 Megabit chip could be arranged into 32-bit words, numbered from 0 to 32767. The address is specified as a binary number to the chip using a set of connections to the *address lines* (such a chip would need 15 such connections). Higher bits of the address would be used to specify different chips, so words in different chips with the same on-chip address have different addresses when the higher bits are considered.

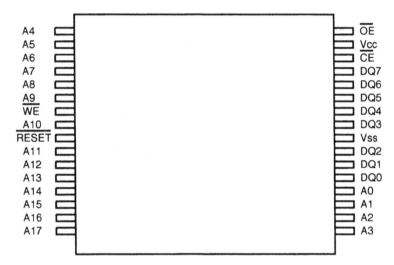

Fig. 1.2   Line drawing of a typical chip. This contains 2 M-bits of memory organised in 8-bit bytes. A0 to A17 specify the 18 address bits necessary, the CE (chip enable) input must be made from the higher address bits using logic circuitry. The OE (output enable) and WE (write enable) connections determine whether the data is read from the chip or written to it.
Vcc and Vss are +5 volts and ground, respectively

So if you could unscramble the connections between the RAM circuitry and the rest of the computer, you would have a set of address connections, a set of data connections, and a few other controls. When the RAM receives a signal on the control connections that tells it that data is being requested, it will set the 32 data connections with the data stored in the 32 ON/OFF transistors that correspond to the word for which the address is given on the address lines.

Another chip you will find somewhere inside the computer is the *Central Processing Unit* or *CPU*. This contains many pieces of circuitry, including memory for two very important words called the *program counter* and the *instruction register*.

These two are the key to the whole computing process. The program counter contains a number which is used as an address. The instruction register tells the CPU of an action to take – such as reading from memory, writing to memory, adding or subtracting or multiplying numbers. These instructions are specified in the *machine code* of the CPU chip concerned: the MMX, the Pentium, the 80486 or 68040, or whatever. So what your computer is actually doing when it is running any program, is running through the following cycle:*

1. Fill the instruction register with the word contained in memory at the address given by the program counter.

2. Carry out the action specified in the instruction register.

3. Increment the program counter so it points to the next instruction.

4. Go back to step 1.

In some CPUs (typically RISC chips, which are reduced-instruction-set chip chips) all instructions are the same length, and the program counter is increased by 1 word (4 bytes) each cycle. This concept offers fewer instructions but what the chip does is done much faster. In others the different instructions occupy different numbers of bytes, and the circuitry has to recognise them, fetch the correct number and adjust the program counter appropriately.

So a program is a long list of machine instructions. Originally it is probably stored on a disk file somewhere, and when you 'run the program' – by typing 'RUN', or the program name, or by double-clicking an icon, or whatever – this list is read from the file into some area of memory. The program counter is then set to the first address of this area of memory (or possibly some other defined *entry point*), and the CPU starts executing the machine instructions of your program.

From the above 4-point list it looks as if the instructions will inevitably all be executed in sequence, one after the other, and then fall off the end. But, although the usual procedure is for each instruction word in memory to follow the next, some of the machine-code instructions that can be executed at step 2 involve altering the program counter. So the flow of control can jump around within your program, skipping over sections or looping back over sections and repeating them. It can even jump out of your area of code to another area, from which it returns to your code later. And finally, when it has finished whatever you set it to do, it will set the program counter to some address corresponding to a resident system program and leave your program area, which is then free to be overwritten by some other program that you may want to run.

You can now put the lid back on the computer, and leave it there. (And turn the machine back on!)

---

* The order of (2) and (3) is interchanged on some systems.

1.3   LANGUAGES AND COMPILERS

A program, when it runs, is a long set of machine-code instructions. These are words of data, and in the bad (?good) old days, these were created by hand. But each instruction achieves very little: even a small program requires several thousand of them. Furthermore the actual instructions vary from one machine type to another, so such software is not transportable.

Very quickly people started to use a more convenient form of writing the code. This was called *Assembler Language* (or *Assembly Language*). A set of alphabetical mnemonics was developed for the machine-code instructions, and for ways of referring to storage locations using symbols rather than the absolute numerical address. Programs are written as alphabetical text.* They are fed into a program called the *assembler*, which outputs the machine-code numerical form. But even the assembler languages suffer from being machine specific, and as each assembler language instruction produces one machine-code instruction, they are still very verbose. Nowadays one uses assembly-language programming only in rare circumstances, for example in connecting a piece of home-made hardware or in an application whose timing is absolutely critical.

Eventually the idea of machine-independent high-level languages developed. Programs can be prepared in statements appropriate for the language, using an ordinary text editor. A program called the *compiler* then takes these statements as input, and produces a corresponding set of machine-code data words as output, and a high-level language statement can produce many machine-code instructions. These are written to a file and can then be loaded into memory and run.

x = a + b * c   high—level language statement

| Assembler equivalent | Machine—language equivalent (Hex) |
Acorn Archimedes	Archimedes RISC processor
LDR R1,c	E59F101C
LDR R2,b	E59F2014
MUL R3,R2,R1	E0030192
LDR R4,a	E59F4008
ADD R3,R3,R4	E0833004
STR R3,x	E58F300C

Fig. 1.3   The machine-language statements equivalent to the assembler and the high-level language. As is the case in all machine-language programs the values in the right-hand column depend on the storage locations used for the variables **a, b, c** and **x**

---

* Using a text editor or indeed, in the old days, on punched cards or paper tape.

Different compilers are written for different machines, so the same high-level language program can be compiled into the correct machine code on different systems. The high-level code is *portable*.

In fact another stage is necessary: a program is not, except in exceptional circumstances, self-sufficient. Your program may do the multiplications and subtractions with single machine-code instructions, but it will need whole sections of code to deal with things like square roots and logarithms. These are available in *libraries* of such standard routines. So the compiled code (an *object module*) has the various supplied routines added to it by a second program, the *link editor*, to produce the complete file of executable code. This then has to be loaded into memory, and control is passed to it. It may be that you have to perform these three stages (compile–link–run) explicitly yourself, or you may have a friendly programming environment which does them all for you at the push of an appropriate key.

It is helpful and usual to distinguish between the outputs from different stages of this process by giving them file names with different extensions. If you write a program called, say, **convert** then the source code would be called **convert.f90** (or **convert.ftn** or **convert.f** or **convert.for** in some systems) if it were written in **Fortran**, and **convert.cc** (or **convert.cpp**) if it were in **C++**. The compiler and linker will produce their machine-readable (and human-unreadable) outputs, of increasing size, with the outputs named appropriately:

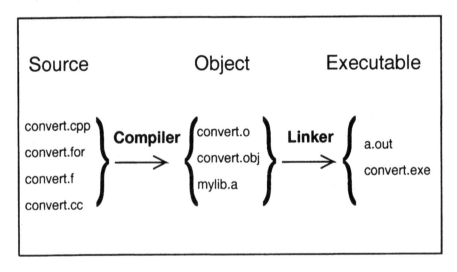

Fig. 1.4   The transformation of the source file to a machine-executable file

Some arrangements are not quite so simple. For example some **Fortran** 'compilers' actually produce output code in **C**, which then has to be compiled to machine code in another step.

1.4   WRITING A SIMPLE PROGRAM

The basic building block of most high-level languages is the *statement*. There are various classes of statement (such as Assignment, Declaration, Input/Output ...) Each statement can be considered as an entity in itself, whereas part of a statement is meaningless. A complete program is a collection of statements.

1.4.1   Writing the source code

The program can be written using a standard text editor, such as a simple word processor, with lines of characters on a page (i.e. a screen), and stored in a file. Such a file is called the *source code* for the program. Usually each language statement corresponds to one line* of the page of text. This helps to make the program readable and understandable by human beings, and that's very important. Writing clear code, whose purpose is apparent to other programmers as well as to oneself, is a programming skill of the utmost value.

In **C++** statements end with the semi-colon ; as a *terminator*. Two or more statements may occur on a line, though doing this tends to make the program harder to read. Conversely, a statement may extend over two or more lines.

In **Fortran** each line is a statement, unless it is continued with the **&** character to incorporate the next line in the statement (etc). The semi-colon ; can be used as *separator* to get two or more statements on a single line.

Comments are used to clarify the program's intentions. These can appear anywhere in the source code: they are ignored by the compiler but are invaluable in making the program readable to a human. In **Fortran** everything on a line after (and including) the ! symbol is treated as a comment and ignored. Likewise in **C++** everything on a line after the double slash // is ignored. This means that comments can be added to the end of any line of code, or may fill a whole line. **C++** also recognises the C-construction whereby code between the /* and */ symbols is treated as a comment, even if this extends over several lines. (This is useful for 'commenting out' sections of code during program debugging.)

`// this is a comment line in C++`	`! this is a comment line in Fortran`
`/* these  two  lines, between the`	`! this & the next  comment  lines each`
`slash—asterisks, are a comment */`	`! need their own exclamation mark`
`x+=2;  y/=2;  // 2 statements  1 line`	`x=x+2;  y=y/2   ! 2 statements  1 line`
`// the next 2  lines contain 1 statement`	`! the next 2  lines contain 1 statement`
`cout  <<  the_best_approximation  <<`	`PRINT * ,  the_best_approximation,  &`
`    " is the best approximation\n";`	`    " is the best approximation"`

---

* Up to 132 characters in **Fortran**.

1.4.2 The program in C++ and Fortran

Let us invent a typical problem – it may even be useful. Given a time interval which is measured in seconds, we want to know what that is in hours, minutes, and seconds. (e.g. 22205 seconds is 6 hours, 10 minutes, and 5 seconds.) Think about how you'd do it on a calculator... it's messy, and requires things like remainders. The computer code is quite simple and the comparison of the two languages gives a flavour of their contrasting styles:

```
//       The C++ version
#include <iostream.h>
main()
{
  // Convert a number of seconds to the
  // "hours, minutes, seconds" notation
  // Written 15–Dec–1996
  // by Alfred User
  int total;
  cout << " Time in seconds? \n";
  cin >> total;          // read in time
  int hour = total/3600; // integer divide
  int minute = (total–3600*hour) / 60;
  int second =
      total–3600*hour–60*minute;
  cout << " A time of " << total
       << " seconds \n";
  cout << " corresponds to ";
  cout << hour << " Hours\n";
  cout << minute << " Minutes and\n";
  cout << second << " Seconds\n";
  return 0;
}
```

```
!        The Fortran 90 version
!
PROGRAM convert
!
!   Convert a number of seconds to the
!   "hours, minutes, seconds" notation
!   Written 15–Dec–1996
!   by Alfred User
INTEGER :: hour, minute, second, total
PRINT * , ' Time in seconds?'
READ * , total
  hour = total/3600    ! 3600 secs/hour
  minute = (total–3600*hour) / 60
  second = &
      total–3600*hour–60*minute
PRINT * , ' A time of ', total , &
  ' seconds'
PRINT * , ' Is equivalent to '
PRINT * , hour , ' Hours'
PRINT * , minute , ' Minutes and'
PRINT * , second , ' Seconds'
STOP ' convert ended normally'
END  PROGRAM convert
```

The programs, in each language, consist of :

- Instructions to the compiler, to tell it that this text is to be converted to a machine-code program. In **C++** (on the left) **main()** tells the compiler that this is a main program, and the curly brackets define its extent. In **Fortran** (on the right) this is done by the **PROGRAM** statement, matched by the final **END** statement, which tells it that the input is complete: there are no more statements in the program. The C++ initial statement, **#include <iostream.h>**, tells the compiler to read a *header file*, **iostream.h**, which is a source file, provided as part of the C++ system, that defines **cout** and **cin**.

- Instructions to the reader (the comments) describing what the program does.

- Statements declaring that some words of memory are to be used for storing integers. These are known within the program as **hour, minute, second** and **total**, but it is important to appreciate that once the compiler has run, there is no name attached to these storage locations, just the address in memory that they correspond to – see Fig. 1.3. In **Fortran** these are all specified at the start of the program, whereas in C++ they may be declared when they are first used.

- A simple **PRINT** statement that sends the text enclosed in quotation marks to the screen. **cout** does the same in C++. (The \n is needed to produce a new line on the output.)

- An instruction to read a number from the keyboard and store it in the location **total**.

- Calculation of the number of hours: as all the numbers are integers, the remainder is dropped.

- Calculation of the number of minutes, i.e. the total number of seconds modulo 60, after taking off the time in complete hours.

- Calculation of the number of seconds, i.e. whatever is left.

- Printing of the final results.

- Concluding statements. The **STOP** statement tells the computer that this program has completed its job and should finish running; the program counter is reset and control is passed back to the operating system. The **return** statement does the same for **C++**. Actually neither of these is necessary as an **END** statement or final bracket causes the program to stop, but this would be casual programming. Also this explicit stop produces the message 'convert ended normally' in **Fortran**, and returns a value of 0 to the operating system in **C++**, which may use it as a test for successful completion.

Problem 1.2   *Write and run the CONVERT program in either* **Fortran** *or* **C++** *on your computer.*

Problem 1.3   *Now do it in the other language. Are the results given exactly the same? Which did you find easier? What happens if you try 604799 seconds – almost 1 week? (You may need to find out about the* **long int** *type in* **C++**.*)*

A proper test would include an extensive range of inputs, such as 12, 1234, 90061 = 1 day, 1 hour, 1 minute and 1 second, 100000, and also values such as −61 and 0.

These initial programs consist of a single procedure – the function **main** in C++ and the main program in **Fortran**. Programs will generally consist of many procedures, all interrelating in a carefully structured way. The same principles of clarity and presentation apply to them all.

1.5   WRITING READABLE PROGRAMS

What is the most important aspect of computer programming? On the basis of our long and often painful experience we are adamant that:

Clarity is the cardinal virtue in computer programming

You may wonder why so much emphasis is placed on this point. After all, if a program does the job it's supposed to, why should anyone care about the style it's written in? If the programmer knows what they've done, why should it matter whether outsiders can comprehend the code? Many programs are written as a one-off job for one's own personal use: who cares what the rest of the world thinks?

But experience shows that most programs have a surprisingly long life, and will require adaptation. In the problem we've invented, the betting is high that in 6 months or so the requirements will change: you may decide the results should include days, for example. In such cases you have to go back to the program and modify it – and now you discover the fallibility of human memory: all the steps which you found obvious at the time of writing, so obvious that you didn't bother to add any comments, will have become totally obscure and baffling. Programming clarity is not only for other people, it is so you can understand your own programs afterwards!

1.5.1   Program "convert"

Even though our program is trivially short, some useful points arise:
- The variables used: **hour**, **minute**, and **second** are all given in lower case. This helps the eye to distinguish them from the standard **Fortran** system words like **PRINT** and **READ**. An important point of difference here: **Fortran** compilers do not distinguish case (upper from lower), while **C++** ones do distinguish case, although by tradition, **C++** programs are almost universally written in lower case.
- The names have been chosen to make their meaning apparent, even at the cost of extra typing: **hr**, **min**, and **sec** would have been easier to write, but harder to read and remember, and **h**, **m**, and **s** even more so.
- The **C++** brackets, { and }, correctly called *braces*, define a *block*: in this case, the main program. They each stand out on a separate line, and the material between them is indented* to make the structure clear.
- It is not a good idea to print out text and a variable of the same name. Confusion arises from statements like **PRINT * , ' answer', answer.**

---

* For better clarity we would prefer to have indented a more significant amount, but we are generally constrained in this book to fit the program into half a page width.

- The program has been given a moderately descriptive name in **Fortran**; there is no choice but **main()** for the main **C++** function.
- There is a sensible use of comment statements.

Let's elaborate on that last point a little:

1. *The overall purpose* of a program is described by a brief comment at the start. At least the reader knows what it's supposed to do.

2. *The author's name* is given. This is not just vanity: the program may well get given to (or taken by!) other users, who can read it off Alfred's disk space, and then get passed further on. When several people collaborate together on a project, writing different parts of it (subprograms) it is even more important to make clear who wrote what and what the precise intention was.

3. *The date of the program* is given, and that's also more important than at first appears. As time goes by, the program will change, and a perennial problem is in establishing which version you've got hold of. A version modified (for any reason) at a later date should have a comment added to say when this happened and what the change consisted of.

4. *Individual statements are usually clear* in their own right, and don't need commenting (except in Assembly Language, where to remain sane you have to comment every statement!) If this is not so, if the logic is tricky or if, as here, numbers are introduced, then a comment is needed to show where they came from.

1.5.2 Comments in source code

While comment statements are useful and important, some people at this point get carried away with enthusiasm, and put so many comments in their code that they obscure the actual program. There is no benefit gained from a comment statement which is itself unclear or unhelpful

```
! Tricky, this bit
```

or wrong

```
! There are 0.0254 inches/metre
```

or from one which repeats information which is clear from the program

```
! print the number x
PRINT *, x              ! number printed
CALL calc               ! perform calculation
```

It is possible to lay out large flashy comments. We find these tend to be unhelpful, detracting attention from the program itself. Actually, one of the most effective devices for making programs clear and comprehensible is the simple blank line (possibly with a // or ! at the start.) These can be used to break up the program into natural units, every 10–20 lines or so, just as sentences are grouped into paragraphs. The break shows where one logical line of thought ends and a new one begins.

1.6 WRITING USABLE PROGRAMS

You may well have taken the points in the previous section to heart and produced a clearly readable program. But there is another aspect:

> **User-friendliness is the second cardinal virtue in computer programming**

Many an excellent program has had its effectiveness ruined by the poor interface it presents to the user. This is not a matter of flashy WIMPS* and menu-buttons, but of the correct overall philosophy. When confronted with a question the human mind can generally envisage a wide range of possible answers – this is particularly true if the user concerned is not a specialist in the area concerned. A program, on the other hand, requires inputs chosen from a fixed and specified range. The matching of these two (incompatible) requirements is the job of the interface and the responsibility for the interface lies with the programmer.

1.6.1 Interacting with the user

Too many programs merely present the user with a question-mark prompt and expect them to be clairvoyant. The key is the sequence

> **Prompt – Check – Recover**

- **Prompt!**
  If the program expects input from the users, it has to say so. You can't expect users to know that they're supposed to type something in at this stage (even if it says so in the instruction manual). They will just sit there wondering why the machine isn't doing anything. In writing the prompt, you have to be specific about what's being asked. A prompt that just says **Input?** or **Option?** is better than nothing, but not by much. The sort of response required should be indicated: **Starting value for x?** obviously requests a number, and **Initial Letter of Surname?** is equally clear. If the response should be one of a choice of options then it helps to give them: **Choose option (SOLVE/CHECK/ERRORS/STOP)?** is reasonably specific, as is **Starting value for x? (0.0 − 25.0)** . If you use perverse, obscure input quantities

---

* Windows, Icons, Mouse, Pull-down-menus ...

because it makes code easier to write, then specify exactly what they have to be (e.g. **Type 1 to continue or 0 to stop**) – though we hope you never sink so low.

- **Check!**
Having got the input, any program should check that it makes sense. Don't just assume that any reasonable person must have entered them sensibly, so you don't need to waste time checking the values. Intelligent people are capable of twisting the logic of what they're supposed to type in, with amazing results. Check that numerical values entered are within the range they should be. In the example given it would be good to check that the number of seconds entered is non-negative, and perhaps that it is more than 59.

- **Recover!**
If the data input doesn't make sense, then the user is presumably under some sort of misapprehension about what's expected, and the program should try and put that right. It's not much help if it just repeats the input message: that gives the user no extra help and they'll probably just repeat the same wrong input.

The program should say there's a problem. (And ring a warning bell or beep if you can!)

Sorry, that doesn't make sense ...

If possible, explain why:

Your chosen option 'EXIT' was not recognised

Give a detailed and helpful explanation about what the input should be:

Your option must be one on the list, typed exactly as given

Now – only now – go back and request the input again:

Choose option (SOLVE/CHECK/ERRORS/STOP)?

User-friendliness also matters for the program output. Just printing a final number, or set of numbers, with no explanation or interpretation is not much help. A statement like **The answer is ...** is not much better.

It is also good practice to echo the input given, especially for numerical values which can easily be mistyped. Even though the values are shown as they are typed in, it may well be that by the end of the program, when the result is printed out and found to be anomalous, the user wonders whether they actually typed their data in correctly. In serious cases the input data can be written to a file which may be stored on disk. The program then reads from the file as if from the keyboard. This removes doubt about what input was used, and is generally safer.

1.7   THE ENVIRONMENT

By this we mean the programming environment for the particular language in use. The range of options is large. At one end of the scale you do it all yourself, writing and changing the program on whatever text editor you choose. You then invoke the compiler by an instruction such as one of:

f90.exe convert.f90   or   **f90 convert.f**   or

tc.exe convert.cpp   or   **cpp convert.cc**

and try to decode the error messages. You can specify *compiler options* at this point, for example to set a high level of optimisation to produce faster code at the expense of longer compilation time, or to say you want to include information for the debugger (§1.11). You may want to specify various input and output files and libraries as well. How this is done varies from system to system; you may have to consult your manual or your local guru.

At the other end you enter an *integrated development environment* – an *IDE* – with a built-in editor, access to all the switches/flags which may be set to modify the compiler behaviour, and all the other features on offer. You strike a key (or use the mouse) and the edited file is sent to the compiler, error messages are listed and decoded, the editor is invoked again but with the cursor pointing to the position in the line which caused the first compiler error, and possibly a suggested correction is offered. After the change is made the cycle repeats. The editor is context sensitive, which means that keywords in the language are colour coded and at any stage clicking on one will bring up definitions, an example or two of the use, and links to related concepts, with which you may have been confused. Some editors colour code everything: strings, integer and real numbers, keywords and the extent of the various constructs (such as **if–then–else**). When such an environment is well-designed and well-crafted it is a delight to use, but be warned! Some apparently offer all this but the designers never seem to have used it and the result is frustration and clumsiness. Perseverance with your IDE is likely to be rewarded with extra efficiency.

1.7.1   MS-DOS and the PC environment

There are several **C++** compilers available for the PC. The environments of Borland and Microsoft are comprehensive, colourful, and in general a pleasure to use. Productivity is enhanced by the use of a properly-designed workspace. Some compilers do not bundle all the software together – the integrated editor is an extra – which is a pity.

The **Fortran 90** list is small at present. One reason may be that the whole standard is so comprehensive that it becomes a huge project. Lahey Fortran offer a full implementation for the PC, as does Salford (NAG), and in addition the **F** compiler (§1.8.1) is available on the PC.

1.7.2   Unix and Linux environments

Unix systems tend to be found on larger-scale workstations – SUN, VAX, etc – but a recent development sees it installed as Linux on a PC, so that virtually any small system may run the unix operating system. The result is amazingly powerful and convenient. **C++/C** is the language of unix and compilers come with the system and are available, free, from GNU. There do not seem to be the IDEs on offer in the unix world that are to be found in the PC world. Unix offers a succession of tools; editing, compiling, debugging, profiling, etc rather than linking them into a monolithic structure, though much of the functionality of an IDE can be done with multi-featured editors such as *emacs* which enable most tasks to be performed without leaving the editor. Unix also provides **make** files and the **make** command which is a powerful way of maintaining the bookkeeping involved in compiling and linking large programs.

The **Fortran 90** unix list is also relatively small. The NAG compiler is fully featured and commercially available, and a Linux F-compiler is freely on offer from Imagine1.

★ 1.8   FUTURE DIRECTIONS OF Fortran AND OF C++

Speculation on the future is always a risky business, but certain trends do seem well-enough established to be worth discussing.

1.8.1   The past, the present and the future of Fortran 90

> *The result is a huge but forgiving language*
> – J. F. Kerrigan: "Migrating to Fortran 90"

Fortran 90 is not just 'another language' but rather the culmination of 40 years of the development of various **Fortrans**, from the original FORmula TRANslation language through **Fortran II**, **Fortran IV**, **Fortran 66**, and **Fortran 77**, amalgamated with all the valuable features of **C++**, and other languages, which could be squeezed in. The array processing, for example, is defined with parallel processing very much in mind. This size and complexity may be a substantial problem for the prospective user. **Fortran 90** is defined in the 1992 ANSI standard, but even 6 years later it is not nearly as common as **C++**. Features that have caused trouble in the past (**COMMON** blocks or arithmetic **IF**s or **EQUIVALENCE** statements etc, etc) for which better alternatives are now known, are still present, but they are marked *obsolescent* and will disappear at the first revision.

The entire **F77** language is present, which means that properly written **F77** code will still run. There are claimed to be millions of lines of such code in production use (but dare one suggest a lot of this is even older than the 1978 standard?) This is a very real practical problem, but eventually there is a danger of preserving such old programs unchanged, partly because all the earlier programmers have left (without leaving adequate notes!) and partly because it is *thought to work*. But if the old code is untestable, not understood and somehow mysteriously suitable for needs 20 years after it was written, then should it be in use at all? If it cannot be verified then isn't it dangerous and even irresponsible to use it?

The goals and achievements of the ANSI committee, X3J3, were a compromise between the preservationists and those who wanted new, modern features. There are many such: modules, the **KIND** parameter, overloaded assignment statements, pointers, free-format source form. Ratification of older extensions include: multiple statements to a line, the ! symbol to initiate a comment at any position, bit-manipulation functions, **INCLUDE** statements, **DO–WHILE** loops, **NAMELIST** input/output, and **IMPLICIT NONE**. In comparison with the **F77** version of **Fortran** 34 statements (keywords, e.g. **DO, CLOSE, END** ...) are unchanged or have a minor change, whereas 52 are either new or have a major change in **Fortran 90**.

Today's programmers should never allow themselves to use the obsolescent constructs included merely for historical compatibility and their use is reprehensible these days. In this book we have avoided them all, but have also made specific choices on the selection of other features. We include those we regard as important, but this is not a comprehensive **Fortran** textbook, only one which draws 50% of its examples from **Fortran**, and so some of the more arcane and esoteric features will not be found here.

Already the trend is clear as to future developments. The next step is the introduction of smaller, modern **Fortrans** which retain the new features but remove clutter.

A standard for **F95** has already appeared and it removes all the obsolescent features and some others. It is described by Metcalf and Reid (1992, 1996a) but as yet no compilers are available. (But some programming discipline and an existing **F90** compiler will work fine.)

The new **F** language (Metcalf and Reid 1996b) is available,* which embraces lower case and eliminates many statements and elements from the **F90** standard.

Lahey[†] have released **elf90** which has broadly the same philosophy, as a subset of their **LF90** full **F90** code. Versions also come free over the Net.

Two of the big advantages of the newer compilers are the reduction in size of the compiler and the guided flexibility given to the programmer.

Of course divergence in these subsets would be disastrous, just like all the **F77** compilers with their useful 'extensions' (such as complex double precision or comments introduced by !). Well, yes, they were certainly useful but it soon made code non-portable and one had the ludicrous situation of having to write clumsy code when elegant and efficient solutions were to hand, but could not be used.

---

* Imagine ltd!! are offering free Linux compilers, and also for other platforms. see info@imagine1.com
† Web address http://www.lahey.com

1.8.2   The future with C++ and C

That **C++** is a more powerful language than is **C** is self-evident. B. Stroustrup (1993, 1997) introduces **C++** as 'essentially a superset of **C**' which allowed the user to define *objects*. These directly relate to the conceptual objects which the program is manipulating. The future of **C++** lies in many ways with the availability and usefulness of defined objects, and many libraries of them have been created, such as linked lists, arrays, character strings, matrices, and graphics classes. Chapter 10 contains more details of object-oriented programming.

★1.9   C++ VERSUS Fortran?

Which of the two languages then would you choose for a safety-critical computing use? You might argue that the more powerful language has better self-checking features (the trusting view), or that it has more in it to go wrong (the cynical view).

You might argue that **C** has been around longer so that more of the potential problems have been discovered. L. Hatton (1995) has written a sobering book which explores these concepts. He points out that even in such a high-profile and well-resourced establishment as the NASA-run Goddard Space Center the detected error rate in writing code is about 6 errors for every 1000 lines of code. The really intriguing result is that it has remained essentially constant over the period 1978–1990, 'dropping' from about 8 at the beginning of the period. Many of the examples given display apparently legal code (i.e. the compiler will not object to it) whose result is undefined. For example the fragment in **C++** and **C**

$$x = a[i] + i++$$

entered with $i = 2$ will result in either **a[2]** + 3 (presumably this was intended by the programmer?) or **a[3]** + 3, *unpredictably* as the compiler has the flexibility to choose the order in which $i++$ and **a[i]** are evaluated. The main conclusion is that certain allowed features of **C++/C** should *never* be used by responsible programmers. Hatton discusses this intriguing concept in detail, and we recommend his arguments to all serious programmers.

Although **Fortran** (with every variable properly typed) may be safer in some ways – a good compiler is more likely to disallow such statements – similar remarks and conclusions do apply to it, and of course to languages in general.

Problems and confusions in the thinking process contribute greatly to such programming errors. We call attention to many of these, such as the mishandling of the logic in the **if–then–else** construct, the comparison of floating-point variables for equality, the reliance on an assumed operator precedence within a single statement (e.g. **a[k++]** = **b[k]**), and so on.

It seems to us that eternal vigilance of the part of the programmer, the validator of the software, and even end-users is most important, combined with a healthy scepticism, whatever the specific language and the specific compiler may be.

★ 1.10   INTERPRETERS

Another way for a computer to 'understand and obey' a program of statements in a high-level language is an *interpreter*. This is a large, sophisticated program which runs in the memory, reading the statements of your program and taking appropriate actions. At one end of the scale BBC BASIC works this way, as do some other BASICs, while at the other, the very large program environments such as the mathematics packages MATLAB^{TM}, MathCAD^{TM}, Mathematica^{TM}, Reduce^{TM}, and others, are also all interpreted. An interpreter encountering a high-level statement such as $x = a + b * c$ does not generate a machine-code equivalent. It consults an internal table which tells it that it knows about variables called **a, b** and **c**, extracts the appropriate numbers, performs the computation, and inserts the result in a location corresponding to a variable known as **x**.

The interpretation method has the big advantage over compilation that the interpreter program is always in control: if anything unexpected is encountered when processing (suppose **b** in the above expression had not been previously defined) then it can pause, give the programmer a complete description of the problem, and any other information they may require (like the values of **a** and **c**), allow them to take remedial action, and carry on. By contrast, a compiled program will typically just crash in an unpredictable way, leaving the programmer the job of guessing/deducing what the cause of the problem may have been, correcting the original program, and going again through the compilation, linking, and running stages. An interpreter can give tremendous efficiency in the development of analysis programs.

It has the disadvantage that your program will run much more slowly, as every line has to be read and parsed before its instructions are carried out. This increases the run-time by a large factor, but 'large' is relative and as always it depends on the circumstances.

Traditionally, languages like BASIC were intended as easy-to-learn languages for beginners, where good diagnostics were necessary but no one really cared about speed, and such languages were introduced for these purposes and were interpreted, whereas languages like **Fortran** were meant for experts in number-crunching applications where speed mattered, and these were compiled. The distinction has later been blurred as people started writing serious programs in BASIC, and thus BASIC compilers came on the market, and at the same time people wanted to learn **Fortran** easily and interpreters for **Fortran** appeared.

★ 1.11   DEBUGGERS

A *debugger* is a system which enables a compiled program to run in a manner similar to an interpreted program. During the compilation and linking stages the variable names are stored somewhere, and the locations assigned to them noted. Also, in the machine code of the program *break points* are inserted: these are essentially jumps out of the user's program into the debugger program. So the program is started and runs as a compiled program until it reaches a break point, when control is passed to the debugger. This can then be used to examine (and change!) values for the variables, and see what is going on if there are puzzling circumstances. The debugger can then pass control back to the program, resuming execution after the break point, which then continues until another break point is reached.

This technique can be useful in tracking down awkward bugs. The main disadvantage is that the compiled and linked programs take up much more space in memory, because of all the lists of variable names and other such information that has to be stored with them. There is usually a switch on the compiler to let you specify whether or not you want to include the extra debug information.

Such additional pieces of software tend to be useful in a direct relation to the convenience of the interface with the user.

★ 1.12   THE OPERATING SYSTEM

> *An elephant is just a mouse with an operating system*
> – Traditional

Despite the mystique and awe surrounding it, there is nothing special about an operating system. It is just a program – albeit a large and complicated one. It runs in the computer, usually being started automatically when the machine is switched on, and thereafter handles the instructions typed in by the user. So when you type **RUN aprogram** this is actually an input to the operating system program, which reads and checks it (i.e. ensures that **RUN** is a known command, and that the **aprogram.exe** file exists) and takes appropriate action (loading **aprogram** from disc to a spare section of memory, and then passing control to that memory by setting the program counter to the start of that memory area, or some other appropriate entry point). When your program has finished then control returns to the operating system. It does a certain amount of tidying up, marking the memory area used for your program as available for other purposes, and then requests its next command from the keyboard.

Another aspect of the operating system is that it provides subprograms, particularly for dealing with *input* and *output*, that your program needs to use. There is a difference between statements like $x = a + b$, which will be compiled into a directly equivalent set of machine-code instructions, and **PRINT ∗ , x**, which will compile into a set of instructions to load the value of **x** somewhere and then jump to a subroutine within the operating system that handles printing. Why do that? Why not just take the appropriate machine-code instructions that do the output in the subroutine and place them in the compiled code? That would certainly save time and memory space spent in jumping around. But even a simple action like printing a number is actually fraught with a great many complexities that depend on the hardware and the system and the way things are being run: is this number printed on the screen, or on a hardware printer, and if an actual printer, what port is it connected to and what control characters will it expect? It is the business of the operating system to know all about such things and handle them in the way it knows best. An input instruction like cin >> x; passes control to the operating system which inputs the number from the keyboard, including such necessary items as allowing the person typing to backspace and repeat characters they've mistyped, before returning to your program with the requested value. For more advanced I/O methods, like writing data to a disc file, the possible complexities are even greater.

There is another important job of the operating system still to be mentioned. Suppose a program is running happily and something out of the ordinary happens: for example, you press a key on the keyboard. This in fact causes something called an *interrupt*: the 4-point scheme of §1.2.2 actually contains a 5th point which is checking whether any interrupts are set. If so then this causes the value of the program counter to be set to some particular address set up beforehand – the *interrupt vector* – which contains a section of code designed to deal with just this eventuality. It will, in this case, read the value of the key you've typed and store it in some appropriate buffer. It then resets the program counter back to its original value, and the program resumes running as before, all unaware that anything has occurred. The operating system has to provide handlers for all the possible interrupts. Interrupts are actually used a great deal, as they're an efficient and convenient way of using the machine. Suppose you want to send a message to a hardware printer part of the way through your program. You could send the first character, wait till it's printed, send the second, wait till that's done, and so on... That's a lot of time for your CPU to spend sitting there idly waiting for the printer to feed paper, move the print head, and so on. It's so much more effective to pass the lot to an operating system routine which sends the first character and then returns control to your program to do something useful until the printer sends an interrupt to say **finished** which kicks the routine back to life; it sends the second character and then passes control back to your program till the next **finished** interrupt is received, and so on till all are done.

2

Variables and Operators

Variables and constants are first described: their names and (briefly) the nature of the different data types. The assignment statement is straightforward: this is followed by a description of the various operators available and how they can be used to build expressions. A typical program is given as an example.

The basic elements of all our computing are the objects we use (*variables* and *constants*), what we do to them and with them (*operators*), and how we give them values (*assignments*). At the simplest level these objects are arithmetical numbers, and there are several different *types* of these. Later we shall meet more sophisticated varieties of predefined types (alphanumeric characters and logical variables are important examples) which have their own appropriate operators. Then as your programming gets more advanced you can design your own variable types, and the operators that act on them.

2.1 VARIABLES

Variables are quantities in programs that you can manipulate and change. Each variable in the program has a unique name and a specified data type. The variable occupies an amount of memory according to this data type, and it has a specific location in memory, its *address*, by which it is accessible to the programmer.

2.1.1 Naming variables – the rules and some guidelines

The (unavoidable) rules governing the choice of names for variables in programs are similar in **Fortran** and **C++**, though there are some crucial differences. In both languages, names must be created from the alphabet (**a–z, A–Z**) and the digits (**0–9**), together with the underscore (_).

In **Fortran**:
- names can be up to 31 characters in length.
- they must begin with a letter, not a digit or underscore.
- **UPPER-CASE** characters *are the same* as **lower-case** characters.
- names can be anything – there are no reserved words.

In **C++**:
- there is no limit on name length.
- names must begin with a letter or underscore, not a digit.
- **UPPER-CASE** characters are **DiFfErEnT** from **lower-case** characters.
- reserved words (see Table 2.1) must not be used.

Table 2.1   Keywords (reserved words) in **C++** which cannot be used as variable names

asm	auto	bool	break	case	catch
char	class	const	continue	default	delete
do	double	else	enum	extern	false
float	for	friend	goto	if	inline
int	long	new	operator	private	protected
public	register	return	short	signed	sizeof
static	struct	switch	template	this	throw
true	try	typedef	union	unsigned	virtual
void	volatile	while			

Actually **bool, true,** and **false** are only reserved in recent versions of **C++** which include the *logical* type. So you might get away with using them, but not for long.

Problem 2.1 *Of the following names, 12 are legal both in* **C++** *and* **Fortran**, *2 of them are legal in* **Fortran** *but not* **C++**, *and 2 in* **C++** *but not* **Fortran**. *Which? Point out a difference between the two languages' treatment of the 12 legal names.*

Example_of_legal_name_in_either_language   nucleus   _priority   switch
**Nucleus   CD_ROM66   max_array_element   tiny   new   value_count
prob_success   cd_rom   min_element_of_array   FUGUE**

Choosing meaningful and helpful names for your variables is one of the most important aspects of writing readable code. So these rules should be supplemented by a few do's and don'ts – which apply to both languages (and to any other!)

- Do use names that mean something. Resist the temptation to call all reals **x** and all integers **i**. Be descriptive – **cm_to_inches** means so much more than **factor**.
- Don't make names too long; 31 characters is far more than adequate.
- Adopt a consistent policy on the use of upper and lower case.
- Don't overwork your variables by using them for different quantities in different parts of the program, e.g. **temp**, **ans** or **result**.
- Never ever make use of the Fortran syntax rule that all spaces are ignorable by breaking up a name with spaces.
- Use the underscore when you want a name containing two (or more) meaningful parts. For example, if you have a variable **a** which may vary between a lower and an upper bound, these could be called **a_lo** and **a_hi**. Or **min_a** and **max_a**.
- Don't use variable names that are similar to (or even, in **Fortran**, the same as) system functions or other words; it confuses the reader. This is perverse and contradicts our first law of programming: Make it readable!

2.1.2 Declaring variables

Variables should be *declared* before use: in **Fortran** this is very strongly recommended; in **C++** it is compulsory. The most basic form of declaration is just to assert the data type of the variable: *integer* variables (see §3.1) are called **INTEGER** in **Fortran** but have the shorter title **int** in **C++**; standard precision *real numbers* (see §3.2) are called **REAL** in **Fortran** and **float** (short for floating-point) in **C++**. For example:

float x_data;	REAL :: x_data
float q_distance , alpha , beta , zeta;	REAL :: height , area , width
int mindex , minus_kk , top;	INTEGER :: number, n_row , m_col

A *declaration statement* is actually doing two things: at *compile time* is establishes that a variable called **x_data** is going to be used as a floating-point real number, and it ensures that by *run time* there will be 4 bytes (or whatever) of memory used for the variable, and the statements in your code referring to **x_data** will use the address of that memory. In more sophisticated programs these two functions may be elaborated, and they may even be split up.

Variables can be *initialised* at the same time as they are declared, if this is appropriate, by combining the declaration statement with the *assignment* statement.

float q = 32.6 , charge = 1.6e−19;	REAL :: Planck = 6.60E−34
float x = 32.78 , y = 1.234E5;	REAL :: fx = 32.71 , dx = −0.038
int index = 15 , beast = 666;	INTEGER :: index = 11 , beast = 666

Although these declarations and assignments look similar in the two languages there is an important difference. **Fortran** only allows initialisation with *constants*, or *constant expressions* (which includes **PARAMETER**s): the compiler allocates the required storage and fills it with the desired value, so this must be so basic that it's known even before the program starts. In **C++** variables are allocated space on the fly while the program is running, and so *variables* and *variable expressions* can also be used as initialisers.

int i = 10;	INTEGER :: i = 10
int range = 101∗3 −7;	INTEGER :: range = 101∗3 − 7
int j = i + 5; // legal in C++	INTEGER :: j = i + 5 ! illegal in F90
	INTEGER , PARAMETER :: size=10
	INTEGER :: j = size + 5 ! OK

In **Fortran**, type declarations must be grouped together at the start of the program. In **C++** this is not insisted on, provided that the variables are declared before use.

int m;	INTEGER :: kk
m = 33;	kk = 33
float x = 0.5; // legal in−line definition	REAL :: z ! illegal after first executable

Some programmers like to put all their declarations together; it helps separate program design into the two parts of first deciding what the data will consist of, and only then going into the detail of how it will be manipulated. (Object-oriented programming (Chapter 10) implies such a style.) Others urge that you should declare a variable when you first use it. When you're reading a program and want to check on the definition of a variable being used, the former style means you always know where to look, the latter that you probably don't have to look so far. The choice of style is a personal one – but it is a good idea to adopt one or the other and to be consistent.

If, in **Fortran**, you do use a variable name without first declaring it, then the type defaults implicitly to **INTEGER** if the first letter of the name is **i , j , k , l , m** or **n**, and to **REAL** for any other letter: **a** thru **h** and **o** thru **z**. Although the practice of using undeclared variables is a throw-back to the bad old days, totally deprecated by all right-thinking programmers, including the authors, it has led to a climate in which i thru n are generally used as first letters of names for integer variables, and the rest of the alphabet generally used for reals. (We will often adhere to this convention.) To enforce the declaration of all variables in **Fortran** the statement **IMPLICIT NONE** is used at the start of a program, and then any variables that are not explicitly defined due to typing errors or memory lapses will be picked up at compile time. Serious programmers *always* use **IMPLICIT NONE**. Compare the error messages resulting from these programs, which are identical apart from the first statement:

	IMPLICIT NONE
REAL :: x1 = 1.0	REAL :: x1 = 1.0
INTEGER :: i , i_max = 99	INTEGER :: i , i_max = 99
! ...	! ...
nexty = SIN(xl)	nexty = SIN(xl)
DO i = 1 , max_i	DO i = 1 , max_i
! ...	! ...
ENDDO	ENDDO

In the first program the compiler will 'helpfully' create extra variables **xl** and **max_i** (because of the typing mistakes) and when the program runs their contents will be undefined – a potentially catastrophic bug that can be hard to track down. In the second program the compiler will complain that the two rogue variables have not been declared. It will also flag **nexty**, as undeclared (did you notice?) instead of quietly setting it as an integer, which is most unlikely to be intended. Good compilers provide *error* messages, which abort the compilation, and *warnings*, which do not. Both should be carefully heeded.

Declarations can give a lot more information about the variables. There are different modes of representing integers. In **C++** these are **short, int** and **long** and are discussed in §2.2.1 and §3.1. In **Fortran** the **KIND** distinguishes them – see §3.1.2. There are also different modes of representing floating-point numbers which will be described in appropriate sections: §2.2.2 and §3.2. In **C++** this information is given by various extensions to the data type name. In **Fortran** the extra pieces of information are called the *attributes*. These come after the type name but before the variables. To show that they are attributes and not variables, they are separated from the type name and from each other by commas, and from the variables by a double colon (::).*

```
INTEGER , PARAMETER  :: m = 100  ! m is a parameter, value 100
INTEGER :: parameter, n =  10   ! "n" and "parameter" are variables
!  "n " is initially 10, "parameter" is undefined
!  Here the name choice, parameter, is confusing and unwise, but legal
```

In simple cases there is less confusion. The :: can be omitted in declarations like **INTEGER j , k** but not when assignments are made, **INTEGER :: pz = 3** . Many programmers always use the double colon and write **REAL :: x** rather than **REAL x** and we recommend the practice, which we consistently use.

---

* This syntactical complexity is the price paid for not restricting the choice of variable names by having a list of reserved words.

★ 2.1.3   More about declarations

A C++ name is either *local* or *global*, and this depends on *where* the declaration occurs. Global variables are declared in a program file outside all the functions (including **main**) in the file. They can be accessed by any function from anywhere within that file. A variable declared within any block (i.e. a set of statements bounded by braces) is local to that block (which includes further blocks nested within it). It cannot be referred to by code outside the defining braces. For example:

```
#include <iostream.h>
    int n_max = 100;              //global variable, access from all blocks
    main ()
    {                             // start first block
        int i;
        {                         // start second block
            int j = 1234;
            i = j;                // OK
            cout << "n_max, i =" << n_max << " " << i << endl;
        }                         // end of second block
        i = j;                    // Illegal since j has now gone
    }
```

the second assignment statement is illegal because when the compiler meets the brace at the end of the second block, it wipes the slate clean and forgets all about the declarations made in internal blocks.

In a **Fortran** module, variables declared at the start, before the **CONTAINS** statement and the subprograms, are likewise available to all the subprograms in the module, and they are also available to any other program that invokes that module through a **USE** statement. Variables declared within a subprogram are not accessible by other subprograms at the same level, though they are available to internal subprograms – thus there is also a hierarchy of local variables, though it is done through subprograms rather than blocks, and has only two levels.

Global data can be useful – and can be dangerous. When you consider the way a set of functions work together their arguments and return values are clearly set out. If they are also messing about with global variables this makes the program harder to analyse, bugs harder to spot, etc. For this reason some gurus urge you not to use global data. On the other hand, if you have many routines which all refer to some quantity it can be very convenient to make it available as global data, rather than adding it to every argument list – though a thorough analysis should avoid both difficulties.

Problem 2.2   *You have routines* **point(x , y)** *and* **line(x1 , y1 , x2 , y2)** *that will plot points and lines on the screen. You want to be able to apply a constant overall scaling factor so that* **mypoint(x , y)** *plots at* (**x∗scale , y∗scale**). *How would you do this (i) with an extra argument (ii) using global data and (iii) not using global data by means of modules or classes (Chapter 10)?*

A C++ variable may also be *static* or *automatic*. This is specified in the declaration using the keyword **static** or **auto** as desired. (**auto** is the default, so the keyword is virtually never used.) Static data is permanent; automatic data is created when needed and lost afterwards.

All global data is static. It is created and initialised before any executable statements are performed, and it stays in existence throughout the program. Local static data is created, initialised and made available when the declaration is first encountered. (If the declaration is encountered more than once, because it's inside a loop or because it's in a function called more than once, then the second time round has no effect.) Even if you leave a block, and the local name becomes unavailable to you, the data is still there, though you can't get at it unless/until the block is re-entered.

Automatic data is created, like static data, when the declaration is encountered. However when the program leaves the block, any automatic data local to that block is destroyed. If the declaration is encountered again, because it's in a loop or for any other reason, then the data is re-created and reinitialised.

Static or dynamic data can be initialised by adding an assignment to the declaration (§2.1.2). If static data is not explicitly initialised by you, then the compiler initialises it to zero. If automatic data is not explicitly initialised, then it is undefined.

The **Fortran** equivalent of **static** is the **SAVE** attribute. A second call of a subprogram will find the values of the local variables have been lost, unless **SAVE** has been specified, *or* the local variable has been initialised when declared.

One use of static variables is in maintaining data in a function. If you want, for example, to count the number of times a function has been called, then you do this with a static variable. Try this routine

```
void f(void)
 {
   static int c1 = 0;
   auto  int c2 = 0;
   cout << "f called " << ++c1 << endl;
   cout << "f called " << ++c2 << endl;
      /* ... */
 }
```

```
SUBROUTINE static_int
  INTEGER :: c1 = 0   ! behaves as SAVE
  INTEGER , SAVE :: c2 = 0
    c1 = c1 + 1
    c2 = c2 + 1
    PRINT * , ' counters', c1 , c2
  RETURN
END SUBROUTINE static_int
```

and call it several times to see the different behaviour of the counters in C++ (only).

Fortran refers to 'automatic data', which means something a little different from the C++ usage, in that it specifically applies to arrays that are declared dynamically. You do this by declaring the array as **ALLOCATABLE** – that's a statement for the compiler – and then it is allocated by a statement that takes effect at run time.

```
REAL , DIMENSION (:) , ALLOCATABLE :: zdata
READ * , n                                 ! get size
ALLOCATE (zdata(1:n))                      ! allocate
READ * , (zdata(i) , i = 1 , n)
```

This attribute is useful when you intend to keep values in an array, but you don't know at compile time how many values there will be. (And you can't say **REAL :: zdata(n)** – not unless **n** is a parameter – which of course has to be given a value at compile time.) Only arrays can be declared allocatable, and the **ALLOCATE** statement can only be used for such arrays, and for pointers.

In **C++** this dynamic allocation is possible but not quite so transparent and user-friendly. Again, **float a[n];** is only valid if **n** is **const int** because **int** on its own is not enough. (And if **n** is a function argument declared as **const int** this is still not good enough!) It is done using a pointer and the operator **new**.

```
float* zdata;              //   pointer to float
int n; cin >> n;           //   get size wanted
zdata = new float[n];      //   reserve block of n floats
                           //   and zdata points to the start of it
for (int i = 0; i<n; i++)
  {
    cin >> zdata[i];       //   equivalent to cin >> *(zdata+i);
  }
```

The **new** statement is really quite clever as it determines the number of bytes or words needed by looking at the size of the type you give it. **new char[100]** would earmark 100 bytes while **new float[100]** would reserve 400. User-defined types (i.e. structures, classes) can also be used, and the compiler knows how much space is needed. This size information is also available to the programmer by using the **sizeof** keyword: **sizeof(x)** gives the size that the variable **x** – of whatever weird type it may be – takes up in memory. It's not strictly a function because it's evaluated at compile time: the compiler knows (that's its job) what **x** is and how much space gets reserved for it. Hence **sizeof** also works with arrays, as in **sizeof(zdata)**.

If – for space reasons and because it is good practice – you want to give back the space when the work is finished, this is done by **delete** and **DEALLOCATE**

```
delete [] zdata;          |              DEALLOCATE(zdata)
```

and the surplus space is released for other purposes – the exact details depend on the stack and memory management of your system. (The square brackets in the **C++** code are used to specify that it is the array pointed to by **zdata** that is to be deleted. **delete zdata** would delete the pointer but leave the object pointed to in place, but inaccessible. This leaves a very messy situation, and is an easy mistake to make!)

A **C++** variable either has *external linkage* or *internal linkage*. 'External linkage' means that it may be accessed by functions in other files, when these are all combined into one executable by the link editor. All local variables have internal linkage – obviously. For global variables the default linkage is external, except for constants when it's internal. To specify a global variable as only having internal linkage you declare it as **static**; this is a rather confusing piece of **C++** syntax as all global variables are static anyway.

Suppose you have a global variable in one file that you want to use in functions in another file. The definition, initialisation, etc, are specified in your file, but when the compiler operates on the other file it has to know about this global variable. You do this by declaring it (in the second file) as external with the keyword **extern**. (This keyword can also be used to override the default internal linkage of **const** variables.) The statement **extern float g;** tells the compiler that a variable **g** exists, and how to handle references to it in the program. The statement **float g;** or **extern float g = 100;** will cause the variable to be created and initialised. So when you combine two (or more) files with global variables used, then there must be one and only one definition causing the variable to be created, and the other files must all declare it as external, but not create it. If you get this wrong there will be error messages at link time.

The **Fortran** solution is a little easier in that you don't have to supply the compiler with details of each external variable; the single **USE** statement picks up all the information about the (global) variables belonging to the module(s) you're using. If there are name clashes between these variables and ones in your program, this can be got round by more complicated forms of the **USE** statement.

If you declare a local variable that has the same name as a global variable (or as a local variable in an outer block) – which can happen quite easily – then the compiler does the sensible thing and uses the innermost declaration. This is called *name hiding*. In the unlikely event of your wishing to override this you can do so using the global scope operator ::. (There's no way to access a variable in an outer block that is hidden by one in an inner block.)

```
int i = 1 ;                           // global
main()
  {
      cout << i ;                     // prints 1
      int i = 2 ;
      cout << i ;                     // prints 2
      cout << ::i ;                   // prints 1
  }
```

Finally let us mention two rather obscure **C++** storage types: **register** and **volatile**. **register** is much the same as **auto**; in addition the compiler knows that this variable will be used very frequently and that it may speed the program up if it can be kept in one of the CPU registers rather than in RAM. The compiler may or may not honour this request (and most compilers are better at spotting this type of optimisation than most programmers, so your request will probably achieve little). Attempting to take the address of a **register** variable is a clearly irrational act, and if you try it the compiler will disapprove. A **volatile** variable is one which may change even if your program hasn't touched it – this can happen if it's actually linked to a piece of hardware, or if an interrupt handler writes to it. This tells the compiler not to take certain short cuts in optimising the code.

2.2   CONSTANTS

A constant can be *literal* which is an actual value appearing in the program. The statement

len_cm = len_inches * 2.54; | len_cm = len_inches * 2.54

includes the literal constant 2.54. In both languages *exponential notation* can be used, so 2.54 can also be written 2.54e0 or 0.254E1 or even as 254.0e−02. (Upper-case 'E' and lower-case 'e' are equivalent.) Alternatively a constant can be *named* – and this is probably better here –

inches_to_cm = 2.54;   // conversion factor
length_cm = length_inches * inches_to_cm;

as this shows where this number 2.54 comes from. There's nothing worse than trying to update a program full of unexplained numerical values, written by a programmer who has now left the project (or by you yourself two years ago).

If you use a variable to contain a constant value there's a danger that it might be inadvertently overwritten. Despite the apparent inconsistency in referring to a 'constant variable' overwriting is such an obvious danger that a remedy is provided in both languages: in **C++** a variable can be declared as **const**, meaning it can't be overwritten (and so it must be initialised). In **Fortran** the **PARAMETER** attribute is used to show that this is a read-only variable:

const float log2 = 0.69315; | REAL , PARAMETER :: log2 = 0.69315
half_life = mean_life / log2; | half_life = mean_life / log2

Attempts to redefine **log2** will be flagged as errors by either compiler.

The compiler decides on the data type of a literal constant if the programmer does not do so. If it is written as an integer, it will be treated as an integer. If it is written with a decimal point then it will be treated as a floating point, even if what comes after the decimal point is zero, and even if it is null! Thus **17** will be taken as an integer, whereas **17.0** and **17.000** and **17.** will be taken as floating-point numbers.

It is good practice to specify your intention about the type of the constant completely rather than trusting to the compiler to guess what you intended. An example of *poor programming* where this advice is ignored is

volume = 4/3 * pi * pow(r , 3); | volume = 4/3 * pi * r**3

which won't even work because the **4/3** calculation will be done using *integer division* and give a result of 1, then converted into 1.0. You have to force the numbers to be taken as floating point by writing

volume = 4.0/3.0 * pi * pow(r , 3); | volume = 4.0/3.0 * pi * r**3

Problem 2.3   *In the above example, consider the advantages or otherwise of* **const double log2 = ln(2.0);** *and the* **Fortran** *equivalent (if there is one).*

2.2.1 Integers in C++

Depending on your **C++** compiler it may be painfully familiar or entirely new to you that on many machines the type **int** is stored in 16 bits. This is described as *int-16*, and its range of −32768 to 32767 is not enough for sensible mathematics. It may bring confusion: for example, 32767 + 1 is undefined − it may be −32768 for some compilers! For unsigned integers 65535 + 1 gives 0. Life is too short to risk these painful traps! Instead 32-bit integers (*int-32*), which can store up to +2147483647 = 2^{31} − 1, should be used as standard in most sensible programs. In practical problems integers greater than 32767 arise quite naturally, whereas it's very unlikely you'll need to use whole numbers above 2000 million.

The compiler may well take the type **int** as int-16 and **long int** as int-32. But this is *not* guaranteed. A 32-bit processor might use int-32 for both! Having decided to use 32-bit integer variables (unless you *know* that they will *always* be smaller than 32767) you then have to discover how to do this on your particular system. If all else fails, read the manual!

Tables 2.2 and 2.3 illustrate the potential danger. They were produced from:

```
// compute the squares of integers to show problems with int−16 in C++
#include <iostream.h>
main()
{
    unsigned int  k_un;
    signed int    m , m1=180 , m2=259;  // 'signed int' is a synonym for 'int'
    long int      n_l;
    cout << "      unsigned−int    signed−int    long−int " << endl;
    for (k_un=m1 , m=m1 , n_l=m1; k_un < m2; k_un++ , m++ , n_l++)
    {
        cout << k_un <<" "<< k_un*k_un <<" "<< m*m <<" "<< n_l*n_l << endl;
    }
    return 0;
}
```

Table 2.2   Squares of the integers 180 to 184 in **C++** using 16-bit arithmetic

integer	unsigned square 16-bit	signed square 16-bit	long integer square 32-bit
180	32400	32400	32400
181	32761	32761	32761
182	33124	−32412	33124
183	33489	−32047	33489
184	33856	−31680	33856

The value of $182^2 = 33124$ is greater than the maximum int-16-storable value of 32767 and so the result wraps round with this compiler. Similarly in Table 2.3 the unsigned int-16 wraps round at $65536 = 256^2$. Chaos may result in a program if the squares of integers are suddenly negative! The fact that the corresponding signed and unsigned **int**s are exactly zero can throw up some programming surprises (e.g. when summing the inverse squares of the integers in §9.6). To show these effects with 32-bit arithmetic set the range to be 46340 ± 4 ($46340 = 65536 \times 0.7071$) and then 65536 ± 4. The point is that with 16 bits the numbers are so small.

Table 2.3   Squares of the integers 254 to 258 in **C++** using 16-bit arithmetic

integer	unsigned square 16-bit	signed square 16-bit	long integer square 32-bit
254	64516	−1020	64516
255	65025	−511	65025
256	0	0	65536
257	513	513	66049
258	1028	1028	66564

★ 2.2.2   Literal real constants, standard and double precision

There is a significant difference between the two compilers: in **Fortran** a floating-point literal constant, used in an expression on the right-hand side of an assignment statement, in an initialisation, or as a function argument, is assumed to be of single precision (**REAL**), i.e. precise to about 1 in 10^7 with an exponent range of $10^{\pm38}$. In **C++** a literal constant is of type **double**, i.e. precise to about 1 in 10^{15} with an exponent range of $10^{\pm308}$ (§3.2.2). Thus **1.23456789** and **1.2345678987654321** are equivalent(!) in **Fortran** (and stored as $\simeq$ **1.234568**) but different in **C++**.

To force a **Fortran** constant to be double precision you specify it by using the **D** form for the exponent. This is done (but also see §3.2.1) as:

```
DOUBLE PRECISION :: two_thirds_s = 0.666666666666666   ! likely error
DOUBLE PRECISION :: two_thirds_d = 0.666666666666666D0
```

The variable **two_thirds_s** will be a **REAL** value masquerading as a double. This apparently perverse result comes by following the **Fortran** rules, given above. See also §2.4.9. There is no way to recover the lost trailing digits.

To force single precision (type **float**) in **C++** the suffix **F** is used. There are no problems about precision loss here, but getting the constant type correct can save unnecessary implicit type conversions (§2.4.9): these are not usually serious but it's nice to get things right. **C++** needs no equivalent to the **1.234D6** scientific notation for **1.234e6** is double precision anyway; to force a float you would say **1.234e6F**.

2.3   THE ASSIGNMENT STATEMENT

The assignment statement is one of the real workhorses of programming. In both languages it has the general form

$$variable = expression$$

The compiler evaluates *expression* – whatever it is – and stores the result in the variable on the left-hand side. For example

jflag = 2;	jflag = 2
quincunx = pow(2 , 15) − 1;	quincunx = 2**15 − 1
circle = 2.0 * pi * r;	circle = 2.0 * pi * r
side = sqrt(area);	side = SQRT(area)

The use of the equals sign for assignment is a little misleading; although the two sides of the equation are equal after the assignment, **a** = **b** is a very different statement from **b** = **a**. Some languages show this by using a symbol which is not symmetric; Pascal and Algol use the double symbol := meant to represent a left-pointing arrow. APL (which works left to right) uses the specific → symbol.

Many assignment statements are used to modify the value of a variable, as in

$$x = x + 1;$$

Because this is so common, **C++** (but not **Fortran**) has the neat shorthand

$$x += 1;$$

This saves you time, particularly if the variable involved has a long name. It may also save the computer time, depending on the compiler, if the variable is an array element whose index is given by a complicated expression. Compare

$$census[3600*hr+60*min+sec] = census[3600*hr+60*min+sec] + count;$$

with

$$census[3600*hr+60*min+sec] += count;$$

All the **C++** arithmetic and bit-manipulation operators (§2.4.2 and §2.4.5) can be joined to an assignment operator in this way to produce:

p += q;	means	p = p + q;		p *= q;	means	p = p * q;
p -= q;	means	p = p - q;		p /= q;	means	p = p / q;

A few spaces can often make a complicated assignment statement much more readable. Compare

$$average=average+(value-average)/count$$

with

$$average = average + (value - average) / count$$

It helps the eye to have the left- and right-hand sides clearly separated, particularly in densely-written pieces of code.

2.4 OPERATORS

The expression on the right-hand side of an assignment can be a simple constant, or a variable, or a function. Or it can be various combinations of all these using operators and, sometimes, parentheses.*

There is an important difference in philosophy between the two languages. An operator in **Fortran** is the same sort of animal as it is in mathematics: it acts on one or more expressions to give a value as a result. It has clear input(s) and a clear output. In **C++** an operator may *also* change one of the 'inputs' as a side-effect. Assignment is an example. In **Fortran** the assignment symbol = is syntactically totally different from, say, the addition operator +. In **C++** assignment and addition are *both* operators: anywhere you put **x** + 2 you can legally put **x** = 2. This is an expression with the value (which you usually don't want) of 2 and a side-effect of setting x equal to 2. In a particular piece of program you may use the result and ignore the side-effect, or use the side-effect and ignore the result, or use both.

Thus the statement **y** = **z** = **137.18;** is equivalent to **y** = (**z** = **137.18**); and also (**y** = (**z** = **137.18**)); It is evaluated in two stages:

(1)   (**z** = **137.18**); is an expression, and which sets **z** to 137.18 as the 'side-effect', and has the value 137.18. So the total expression is

(2)   (**y** = **137.18**); which is *also* an expression having the value 137.18 and which sets **y** to 137.18 as its 'side-effect'.

2.4.1 Monadic and dyadic operators

There are two major classes of operators. *Monadic* or *unary operators* act on one argument, and *dyadic* or *binary operators*† act on two. (**C++** also has a single example of an operator with three arguments, the *conditional operator* **p?q:r** §4.4.1.)

For example in **C++** the tilde symbol ~ indicates a monadic operator which returns the ones-complement of its argument. The division operator / is dyadic since **a/b** has two arguments, **a** and **b**.

To confuse this neat picture we now have to tell you that some operator symbols can denote monadic *or* dyadic operators. Sometimes the monadic and dyadic versions are completely different, sometimes they're similar. In both languages the simple plus and minus operators can be used in a monadic or dyadic way. The difference between the two forms is not usually a worry.

```
a = b + c;          a = b + c ! dyadic   : addition
a = +d;             a = +d    ! monadic : does nothing
e = f − g;          e = f − g ! dyadic   : subtraction
e = −h;             e = −h    ! monadic : negation
```

---

* The term 'brackets' is ambiguous. Where it is necessary to distinguish between the different varieties we will refer to (parentheses), { braces },[square brackets] and < angular brackets >.

† This is a rather confusing name, as 'binary' operators have no necessary connection with 'binary' numbers. This is why we prefer to use the terms monadic and dyadic.

In **Fortran** this is all there is to it. But in C++, which has many more operators, the asterisk * and the ampersand & symbols each have very different meanings when used monadically and dyadically:

```
a = b * c;  // dyadic : multiplication   a = b & c;  // dyadic : bitwise AND
x = *ptr;   // monadic: contents–of      ptr = &x;   // monadic : address–of
```

The contents-of and address-of operators refer to pointers, which are described in §3.6, and in detail in Chapter 7.

2.4.2 Simple arithmetical operators

In both languages addition, subtraction, and multiplication, represented by +, −, and *, present no problems. The only tricky point is the division of two integers: this gives a result which is also an integer, and any remainder is discarded. Thus **41/10** gives **4**. So does **49/10**.

Sometimes you want to know the remainder after an integer division **i/j**. The value of **i − j*(i/j)** always gives this remainder but is messy to code. Instead C++ provides the neat **%** operator: **i % j** gives the remainder you want. In **Fortran** a function is provided: **MOD(i , j)** returns the desired remainder.

It all sounds easy, doesn't it? The fun starts when one of the integers is negative. What do you think the value of **(−10)/3** or **10/(−3)** should be? **Fortran** is consistent: the rounding is done *towards zero* so positive results are rounded down and negative results are rounded up. This will produce negative remainders. There is an alternative function **MODULO** which gives the true modulus, i.e. it is always positive.

The direction of rounding with negative integers in C++ is not defined; it may be towards zero or away from zero. Presumably this is because different computer architectures may adopt different systems, and to impose one system on another would give considerable overheads. This pushes the problem on to the programmer. If you do have to work with negative-integer division, then remember that whatever works on your machine is not guaranteed on any other! The only safe option is to handle it yourself or to use a more defined language. Or rephrase the problem. What *is* guaranteed is:

In C++ for integers **a , b** with **(b != 0)**

$$(a/b)*b + a\%b \equiv a$$

In **Fortran** for integers **p , q** with **(q /= 0)**

$$(p/q)*q + MOD(p , q) \equiv p$$

Problem 2.4   *Calculate (-10)/3 and 10/(-3) and some other divisions with negative numbers on your system using (if possible) both languages. Check the results are consistent and predictable.*

2.4.3 Auto incrementing

Further helpful features of **C++**, absent from **Fortran**, are the autoincrementing operators ++ and −−. These are monadic operators which can come before or after the argument they apply to. In both cases they change the value of the argument (which must be a variable) by 1, incrementing for ++ and decrementing for −−. If the operator precedes the variable it returns the value *after* the change; if it comes after then it returns the value *before* the change was made.

```
index  = 1;
j_list = index++;        //  set j_list to 1, then set  index  to 2
next   = ++index;        //  set index  to 3, then set  next   to 3
```

These are particularly useful in loops, and in truth give a distinctive look to **C++/C** programs. Section §4.7 gives many examples.

★ 2.4.4 Exponentiation

Although there are many more operators in **C++** than in **Fortran**, the exponentiation operator, ∗∗ in **Fortran**, is noteworthy for its absence. Instead there is an function **pow** which does the job, although not so elegantly. (To use it you have to put **#include <math.h>** at the start of your program.)

```
area = pi * pow(r, 2);    // C++    |    area = pi * r**2    ! F90
```

The designers of the original **C** language, presumably following Pascal, left out the exponentiation operator, deciding it was not useful enough in day-to-day computing to merit a specific operator, and that it was liable to abuse by ignorant users. From our viewpoint this is unfortunate (since exponentiation is common in scientific programs, and anyway none of our readers is ignorant), but we have to live with it.

In either language (indeed, in any language) exponentiation has to be used carefully. It is not a hardware feature in most CPU chips, and a^b is probably evaluated as $e^{b \ln a}$, i.e. by

(i) finding the logarithm of a in a look-up table (SLOW)

(ii) multiplying by b and

(iii) looking up the exponential of the result in another look-up table (SLOW).

If b happens to have the value 0.763, then this calculation is necessary, but if b happens to have the value 2, then the computation would be done more quickly and more accurately as $a \times a$.

The C++ function **pow** has a prototype* **double pow(double fbase , double fpower)** (in our system) and so all variables are cast to type **double** before use and hence **pow(1.6 , 2)** and **pow(1.6 , 2.0)** will be treated identically.

Fortran is aware of what is requested; if you code **a**2** then the compiler will implement the calculation as **a*a**, but if you code **a**2.0** it will do it the slow way. Similarly with square roots. The intrinsic function **sqrt(x)** is very fast, efficient and accurate, whereas **x**0.5** uses the general method.

> **Problem 2.5** *Investigate the speed, and accuracy, in* **Fortran** *with which* **a*a, a**2, a**2.0** *and* **a**1.999** *are calculated on your system. Try also* **a**(0.5)** *and compare with* **sqrt(a)** *(if you are using C++ translate as appropriate). If possible, try it with different compilers, or different compiler optimisation levels. To get sensible timing you may need to repeat the calculation, say, 10000 times.*

Another point to watch with exponentiation in C++ and **Fortran** is that errors arise whenever **a** < 0 and **b** is not an integer. If we take **a** = −1.0 and **b** = 0.5 then **pow(a , b)** or **a**b** is $\sqrt{-1} = 0.0 + 1.0\,i$ which is a *complex* number. Both languages reject the attempted calculation by giving a run-time error. (**Fortran** can even spot erroneous expressions like **(−1.0)**0.5** at compile time.)

Both languages support complex numbers and all the above calculations are *valid* when **a** is complex, for all **b** values, integer, real and complex. Thus if **z** = **complex(0.0 , −1.0)** then **pow(z , 1./3.) = (0.5 , 0.866025)** corresponding to $i^{1/3} = (1 + \sqrt{3})/2$.

If the value of **b** is an integer then the expression **(−1.0)**b** is real: if **a** = −1.0 **b** = −18 then **a**b** is just +1.0. But the compiler may not realise this if **b** is floating point! **Fortran** rejects any real power of a negative number, even **(−1.0)**2.0** and the tricky **(−1.0)**0.0**, although **(−1.0)**2** is fine. (**−2.0**0.5** also runs without giving an error, but it gives a result of −1.4142.. because the compiler's rules of operator precedence mean this is taken as −(2.0**0.5) – see §2.4.6.) C++ is slightly different, accepting **pow(−1.0 , 2.000)** on the grounds that 2.000 is an integer while correctly rejecting **pow(−1.0 , 2.001)**. It sets **pow(0.0 , 0.5)** to 1, according to the rules of algebra and, rather more arbitrarily, **pow(0.0 , 0.0)** also evaluates to 1. In **Fortran** even **(−1.0)**1.0** or **(−1.0)**0** can lead to run-time error messages.

Wrong thinking leads to wrong programming. These examples will arise when formulae like $(-1.0)^k$ are being evaluated for various integer k-values, say $k = 0, 1, 2, 3 \ldots 50$. Thus the real question is not 'what is $(-1.0)^k$?' but rather 'is k even or odd?' and return +1.0 or −1.0 accordingly.

> **Problem 2.6** *Write a function or expression to tell if an integer is even or odd.*

---

* This prototype in **math.h** tells the compiler details about the named function; that it is of type double and takes two arguments, also of type double.

2.4.5 Bit operators and bit functions

For 'sordid' operations that act on a word or words on a bit-by-bit basis, **C++** provides a set of *bit operators* and **Fortran** a set of *bit functions*.

The operators in **C++** are AND (**&**), inclusive OR (|), exclusive OR (^) and NOT (~). The operators $<<$ and $>>$ shift the word to left or right a given number (say p) places. For a left shift, $a << p$, zeros are added at the right as required. For a right shift, $a >> p$, where a is of type **int**, then it is machine dependent (!) whether zeros are added at the left or whether the most significant bit is repeated. If the most significant bit is zero then the result is the same, if not then this can bring trouble. If $a \geq 0$, or if unsigned integers are used, zeros are added on the left. The reason for the ambiguity is that propagating the top bit is the correct thing to do if you're doing arithmetic operations on twos-complement negative numbers; if this is not what you are doing then use **unsigned int**.

Thus, in octal, using 16-bit words we have the following example:

Table 2.4   Calculations in **C++** with bit operators in octal with 16-bit words

Bit Operations			Result
~012345			365432
000777	&	123456	000456
000555			123456	123557
000555	^	123456	123103
123456	<<	3	234560
023456	>>	6	000234
123456	>>	6	001234 or 177234

In **Fortran**, functions are used. **IAND(i , j)**, **IOR(i , j)**, **IEOR(i , j)** and **NOT(i)** return what you'd expect. (The **I** at the front is because they give integer values, and are named following **Fortran** traditions.) Only one shift function **ISHIFT(i , n)** is needed: it shifts **i** by **n** bits, and **n** can be positive (for left shift) or negative (for right shift). **Fortran** specifies that zeros are added at the appropriate end. There is another function **ISHIFTC** which does a circular shift, moving the bits from the end they are dropped off and inserting them at the other end. **Fortran** also provides functions **IBSET(word , position)** and **IBCLR(word , position)** which set and clear particular bits in a word; in **C++** such actions have to be done 'by hand' using the inclusive-OR and AND operators.

Problem 2.7   *Check the above calculations, particularly the last one, on your system. If possible try using different compilers.*

Bit numbering is an area where differences in user definitions can lead to mistakes and misunderstandings. In the **Fortran** functions the bits are numbered from right to left beginning with 0: thus bit 7 is the most significant bit of a byte, and the most significant bit of a full word is bit 31.

2.4.6   Operator precedence

In an expression containing two (or more) operators, it may be important to know which gets evaluated first. This is just like elementary school arithmetic, when we were taught how $4 + 2 \times 3$ is 10, not 18, as 'multiplication and division come before addition and subtraction'. So in both languages **4+2*3** will multiply 2 and 3 and then add 4 to the result, giving 10. We say that the *precedence* of the multiplication operator * is higher than that of the addition operator +. Similarly **1.0+x/y** and **b*b−4.0*a*c** are treated as you would expect. The use of parentheses will override the operator precedence if that's what you want to do: **(4+2)*3** would give 18.

But these computing languages contain many more than the simple 4 arithmetic operators, so **Fortran** has not just 2 levels of precedence but 10. These are shown in Table 2.5.

Table 2.5   Operators in **Fortran**

Precedence	Symbol	Arguments	Name
1	**	Dyadic	Exponentiation
2	* /	Dyadic	Multiplication and Division
3	+ −	Monadic	Identity and Negation
4	+ −	Dyadic	Addition and Subtraction
5	//	Dyadic	Concatenation
6	== / = < <= > >=	Dyadic	Equality and Relational
7	.NOT.	Monadic	Logical Negation
8	.AND.	Dyadic	Logical And
9	.OR.	Dyadic	Logical Or
10	.EQV. .NEQV.	Dyadic	Logical Equivalence

Problem 2.8   *What are the values given by the integer operations*
(a) **1+2−3**   *(b)* **1+2/3**   *(c)* **−1 −2**   *(d)* **−1 +2**   *(e)* **3+4 −5**
(f) **1+2**3**   *(g)* **2**3+1**   *(h)* **(2**3)+1**   *and*   *(i)* **2**(3+1)***?*

The **C++** version of this table (Table 2.6) is much more complicated. Having more operators in the basic language makes it more powerful, at the cost of the complexity which makes it harder to remember what they all do.

Table 2.6   Operators in C++

Precedence	Symbol	Arguments	Name	Section
1	::	Dyadic	Scope Resolution	10.1.1
	::	Monadic	Global	2.1.3
2	. − >	Dyadic	Member Selection	3.9
	[ ]	Dyadic	Subscript Brackets	3.10
	()	Dyadic	Function Argument Brackets	5.2
	+ + − −	Monadic	Post Increment and Post Decrement	2.4.3
3	+ + − −	Monadic	Pre Increment and Pre Decrement	2.4.3
	+ −	Monadic	Identity and Negation	2.4.1
	& *	Monadic	Address-of and Contents-of	3.6
	~ !	Monadic	Bitwise and Logical Negation	2.4.5
	new delete	Monadic	Storage Management	2.1.3
	sizeof	Monadic	Storage Management	2.1.3
4	. * − > *	Dyadic	Member Function Selection	7.2 9
5	* / %	Dyadic	Multiply Divide Remainder	2.4.2
6	+ −	Dyadic	Addition Subtraction	2.4.1
7	<< >>	Dyadic	Shift	2.4.5
8	< <= > >=	Dyadic	Relational	4.1.3
9	== !=	Dyadic	Equality Inequality	4.1.3
10	&	Dyadic	Bitwise AND	2.4.5
11	^	Dyadic	Bitwise Exclusive OR	2.4.5
12	|	Dyadic	Bitwise Inclusive OR	2.4.5
13	&&	Dyadic	Logical AND	4.1.2
14	||	Dyadic	Logical Inclusive OR	4.1.2
15	?:	Triadic	Conditional Operator	4.4.1
16	= += etc.	Dyadic	Assignment	2.3
17	**throw**	Monadic	Throw Exception	4.8.1
18	,	Dyadic	Sequencing	2.4.8

This is not a table to memorise. Indeed, it's not really a table to use! It's often safer and easier to read if you use parentheses to spell out your intentions, rather than relying on the precedence table. For one thing, it's quite easy to misread the table and confuse the monadic and dyadic versions of the same operator.

Problem 2.9   *If* $i = 1$, $j = 2$, *and* $k = 3$, *what are the values given by*
(a) i+j −k   *(b)* i+ ++j   *(c)* i++ + j   *(d)* i/j −k
(e) −i −j −k   *(f)* i ++ + ++ j   *(g)* i+++j?

2.4.7 Operator associativity

If two operators with the same precedence (which may be the same operator twice) occur in an expression, then you may need to know which is evaluated first. Will **a • b • c** be regarded as **a • (b • c)** (*association to the right*) or as **(a • b) • c** (*association to the left*)? For some operators (like + and *) this doesn't matter, but for others (like − and /) it does. This is determined by whether the operator *associates* to the left or right. That is to say: parentheses are inserted starting from the left or the right as appropriate. All operators of equal precedence must associate in the same way, otherwise the value would not be well defined.

In both languages *most* operators associate to the left, so

$$a - b + c - d + e - f$$

is equivalent to

$$((((a - b) + c) - d) + e) - f$$

with parentheses inserted starting from the left.

In **Fortran** only the exponentiation operator ** is right-associating. **a**b**c**d** is taken as **a**(b**(c**d)))**, in accord with everyday (?) usage of quantities like x^{y^z}.

For **C++** the exceptions are

  (i) the monadic operators
  (ii) the triadic conditional operator, and
  (iii) assignment operators.

Again, this is meant to correspond to what people are naturally used to.

Problem 2.10   *Explain why* **a = b = c = 1.0;** *sets all 3 variables to 1.0 This is a very useful thing to do in* **C++** *and is entirely within the syntax.*

Problem 2.11   *What are the values given by*
  (a) **2./1./3.**   *(b)* **2/1/3**   *(c)* **4./3.*2.**   *(d)* **2**3**4** *?*

2.4.8 Operator ambiguities

The logical order of evaluation and the actual order may not be the same. In a sequence like **a = x+y+z** the compiler is free to decide whether to add **x** to **y**, and then add **z** to the intermediate result, or to form **y+z** and then add **x**. Why should this matter? Suppose a previous line contains, say **b = u+y+z**, a clever compiler would evaluate the intermediate **y+z** only once, keeping it in a spare register, which would help the program run faster. Indeed, **Fortran** is allowed (like the mail order companies) to substitute another expression of equal value if it sees good reason. **C++** has the same freedom under the 'as-if' rule. For many CPUs multiplication is much quicker than division, and on such a system a compiler might evaluate **x/y/z** as **x/(y*z)**, and **x/10.0** might be replaced by **x*0.1**.

If this flexibility gives problems you can override it by using parentheses: $(x/y)/z$ will be evaluated using two divisions. (You might want to do this if, for example, x, y and z were all small, such that $y*z$ was so small it would be rounded to zero.)

Most of the time this goes on without the programmer knowing or caring. But if the objects being calculated have side-effects, something that happens with operators in C++ and with functions in both C++ and **Fortran**, then the actual order may matter. For example in C++ if i has the value 0, then $j = i + i++;$, we would expect, could assign either 0 or 1 to j, depending on whether the first part of the expression (i) is evaluated before the second part ($i++$) or afterwards. In fact the result could be anything at all. Likewise in **Fortran** $a = f(x)+x$ is ambiguous if the function f modifies its argument.

Such expressions are clearly very dangerous: in C++ the result is undefined, so that what happens with one compiler may happen differently with another compiler, or even with a new version of the same compiler. In **Fortran**, which as usual leaves less scope for the programmer to do themselves mischief, they are illegal and the compiler should reject them.

In C++ compound logical expressions (which use the logical AND and logical inclusive OR operators && and || – to be discussed fully in §4.1.2) are a special case. They are evaluated left to right, and if the result is obvious then the evaluation stops. If a is false then the result of a && b is bound to be false, whatever the value of b. So the compiler can, and does, save time (maybe quite a lot of time if b is replaced by something long and complicated, such as a function call) by evaluating the second expression only if the first is true. It can also help your program logic if b is contingent on a, for example a could be a logical function that establishes the existence or otherwise of a file, and b a second function which opens it for input. **Fortran** takes the opposite view: in the expression a **.AND.** b it may evaluate either a or b first, and use that freedom to avoid evaluation of other parts of the expression once the overall value is inevitable.

Problem 2.12   *Write a* **Fortran** *program with a sequence like*
```
IF ( a>0 ) THEN
    IF ( a<0 .OR. f(x)>y ) THEN
    ENDIF
ENDIF
    ! ...
FUNCTION f(x)
    ! ...
    PRINT * , ' f(x) called with x = ' , x
    ! ...
END FUNCTION f
```
and investigate whether $f(x)$ *is called or not. Try different logical expressions. Can you control your compiler?*

Another exception is a convenient **C++** feature called the *comma operator* or *sequencing operator*

expression_1 , expression_2

Its definition is at first surprising: **a , b** evaluates the expression **a**, throws the resulting value away(!), and gives the value of the expression **b**. Thus **s = 1 , s + 2** returns the value 3, and sets **s** to 1 as a side-effect. Clearly, by its very nature, the first expression must be evaluated before the second, and this is indeed the case. If the expression occurs in an environment where it may get confused with other commas, for example those separating arguments in a function call, then parentheses have to be used; it may help clarity to use them anyway (and the comma operator has the lowest possible precedence). The comma-expression can be useful when you want to carry out two operations in a place where the syntax only allows you to do one, for example in the initialisation step of a **for** loop

```
for (sum=0.0 , i=100;  i<201;  i++)  // sum contents 100– 200 of the array
           sum  +=  array[i];
```

2.4.9   Type conversion

If you assign an expression of one type to a variable of another type then this may have unexpected consequences. Statements like **x = 1** where an integer value is assigned to a real variable **x** do not give any trouble, as any compiler will spot this and convert the **1** from integer to floating-point representation. But statements like **j = 1.8**, where a floating-point real number is assigned to an integer, are more tricky. What the compiler (both in **Fortran** and **C++**) does in this case is round off the part beyond the decimal point, and store the integer part (here **1**) in the integer variable. When an operator is applied to two variables of different types, it may be necessary to convert one of them to match the other. An expression like **i + x** will require the conversion of the integer **i** to floating-point format before it can be added to the floating-point variable **x**. Less obviously, adding two integers of different lengths, or adding a single and a double-precision floating-point number, also require one of these *implicit conversions*. The rules by which the compiler decides what to do are quite complicated, but they are generally sensible and you're unlikely to need to know details.

In many cases you want to force an *explicit conversion* from one type to another. This is done by specific functions with names identical to the type desired. This means that the code may look similar to the following:

```
int     jmin;                    INTEGER :: jmax
float   xmin;                    REAL      :: xmax
     /* ... */                        ! ...
xmin  =  float(jmin);   // cast   xmax  =  REAL(jmax)    ! function
```

The use of **REAL** or **float** helps the overall clarity of a program by demonstrating the programmer's intentions (although the programs would run equally well without them). It is vital when you want to divide two integers and get a real number.

Problem 2.13   *Which of the following give the value 3, and why?*
(a) 22/7   *(b)* 22.0/7.0   *(c)* float(22)/float(7)   *(d)* float(22/7)
(e) float(22)/7   *(f)* 22/float(7)   *(g)* 22.F/3   *(h)* 355.F/113.F
(If using Fortran *replace* float *by* REAL.*)*

Remember that converting positive real numbers to integers rounds down by omitting the leftover fraction. **Fortran** converts negative real numbers to integers by rounding up; what **C++** does is undefined and unpredictable, but of course, consistent! See §2.4.2.

Careful programmers will tend to avoid implicit conversion. They will always write (if **x** is a floating-point variable) **x = 1.0** rather than **x = 1**, because it avoids a type conversion at run time. In **C++** they would write **float x = 1.0F;** rather than **float x = 1.0;** because even the latter apparently innocent statement includes a double-to-float conversion (as the default floating-point constant type is double in **C++**). For an integer **i** one should write **i = 99** rather than **i = 99.0** out of a general distrust for floating-point-to-integer conversions, even in a case like this where the result is not affected by integer rounding. If there has to be a conversion it is better to write **i = int(x);** or **x = REAL(i)** rather than **i = x;** and **x = i** as, even though these probably compile to the same code, it makes it clear to anyone reading the program that a conversion is occurring. Even this is not enough for it leads us into a subtle trap. What would you expect as the outcome of the following code?

```
cout << "int_double = " << int(2.5/0.1);      PRINT '(I4)' , INT(2.4D0/0.1D0)
cout << "int_float  = " << int(2.5F/0.1F);    PRINT '(I4)' , INT(2.1/0.1)
```

You may expect 25, 24 and 21, but you get 24, 23 and 20. Each line gives the integer equivalent as 1 less than it 'should' be! If you can explain why you have a good working understanding of computer reals and integers and will appreciate why correct code has the form

$$\text{int (computed real} + 0.5)$$

This is also why we never compare computed reals for equality. A result of 24.99999 is *rounded down* to 24 and the extra 0.5 is to compensate for such inevitable errors of floating-point arithmetic. **Fortran** provides a useful addition to **INT()** (which rounds down towards zero) in **NINT()**, which rounds to the *nearest integer*. Its use removes the problem in the code above. You will find with

```
PRINT '(2I4)' , INT(2.4D0/0.1D0) , NINT(2.4D0/0.1D0)
PRINT '(2I4)' , INT(2.1/0.1) ,      NINT(2.1/0.1)
```

that 24 and 21 are now correctly returned.

ⵏ 2.4.10 Type casting in C++

C++ also allows conversions using the *cast* notation, which puts the brackets in another way round

$$\text{int } i \; = \; (\text{int}) \; x; \qquad // \text{ same as int } i \; = \; \text{int}(x);$$

though we prefer the *functional* notation **int(x)**. It is true that this can't be used if the type doesn't have a simple name, for example if you want to cast an integer as pointer-to-char so as to access a char at the specific memory location 123456.

```
int i = 123456;
char *p;
p = (char*) i;          // legal (though dangerous)
p = char*(i);           // illegal and nonsense
```

but you can bypass this problem by defining a type using **typedef** (§3.8) which sets the second name as a synonym for the first.

```
typedef char   *Pointer_to_char;
p = Pointer_to_char(i);           //legal
```

Conversions or casts of pointers are a bit different. Converting a float **x** to an integer **i** gives a different bit pattern, which has (insofar as is possible) the same value as the original. Converting a pointer-to-float to a pointer-to-int

```
float x = 1.2345;               // a typical floating—point number
float *p_to_float = &x;         // points to x
int *p_to_int;
p_to_int = (int*) p_to_float;   // pointers now have same value
                                // but different interpretation
int k = *p_to_int;              // k is a very peculiar number
```

gives the same bit-pattern memory address, but changes its value by changing its interpretation. But take care!

2.5 A TYPICAL PROGRAM

Here is a program to illustrate many of the points covered in the chapter. Its purpose is to compare, over a range of n-values, the exact value of the log factorial

$$\ln n! = \ln[n(n-1)(n-3)\cdots 3\cdot 2\cdot 1]$$

with three different approximations: the simple form of Stirling's approximation

$$\ln n! = n \ln n - n$$

the somewhat more accurate estimate

$$\ln n! = n \ln n - n + \ln(\sqrt{2\pi n})$$

and finally including the additional terms

$$+\frac{1}{12n} - \frac{1}{360n^3} + \frac{1}{1260n^5} - \frac{1}{1680n^7} + \frac{1}{1188n^9} \cdots$$

```
// Approximations to log(n!) n>0
#include <iostream.h>
#include <math.h>
main()
{
long int    i , n , int_prod;
float xn , r , r2 , real_prod , logfac;
float fac1 , fac2 , fac3;
for ( ; ; )
  {
  cout << " Integer to test "<<
    "[ n > 0 ] \n";
  cin >> n ; if (n < 1) break;
  xn = float(n); r = 1.0/xn; r2 = r*r;
  int_prod  = 1; real_prod = 1.0;
    logfac  = 0.0;
  for (i = 1; i<=n; i++)
    {
    if (n < 13) int_prod *= i;
    else if (n < 35) real_prod *= float(i);
    else logfac += log(float(i));
    }
  if (n < 13) logfac = log(float(int_prod));
  else if (n<35) logfac = log(real_prod);
  fac1 = xn * log(xn) − xn;
  fac2 = fac1 + 0.5*(log(2.0*M_PI*xn));
  fac3 = fac2 + r/12.* (1. − r2*(1./30.−r2
  *(1./105. −r2*(1./140. −r2/99.))));
  cout << "n , log n! , and "
    << "approximations \n";
  cout << n <<" "<< logfac <<" "<< fac1
  << " " << fac2 << " " << fac3 << endl;
  }
cout << " n < 1 in factorial \n";
return 0;
}
```

```
PROGRAM log_factorial
IMPLICIT NONE
INTEGER :: i , n , int_prod
REAL :: xn , r , r2 , real_prod , PI
REAL :: fac1 , fac2 , fac3 , logfac
PI = 4.d0*ATAN(1.0d0)
DO
PRINT * , 'Integer factorial n > 0'
READ * , n; IF (n < 1) STOP 'n < 1 '
xn = REAL(n); r = 1.0/xn; r2 = r*r
IF (n < 13) THEN
   int_prod = 1
   DO i = 1 , n
     int_prod = int_prod * i
   ENDDO
   logfac = log(REAL(int_prod))
ELSEIF (n < 35) THEN
   real_prod = 1.0
   DO i = 1 , n
     real_prod = real_prod*REAL(i)
   ENDDO
   logfac = log(real_prod)
ELSE
   logfac = 0.0
   DO i = 1 , n
     logfac = logfac + log(REAL(i))
   ENDDO
ENDIF
fac1 = xn * log(xn) − xn
fac2 = fac1 + 0.5*(log(2.0*PI*xn))
fac3 = fac2 + (r/12.)*(1. − r2*(1./30.&
−r2*(1./105.−r2*(1./140.−r2/99.))))
PRINT * , 'n , log n! , & approximations'
PRINT * , n , logfac , fac1 , fac2 , fac3
ENDDO
END PROGRAM log_factorial
```

3

Data Structure

The various ways of representing data in the computer are discussed in detail, starting with integers and floating points, and proceeding to complex numbers. Logical values, characters, and pointers are discussed briefly; full details are in later chapters. Structures are described and explained, though a fuller account of their use as objects is also deferred. Arrays of one or more dimensions are dealt with at some length.

Scientific programming is about manipulating numbers, and collections of numbers are generally known as *data*.

There are several *types* of data. As scientists we are all expected to be numerate, and effortlessly to recognise and appreciate the difference between *integers* (such as 2, 17, -111 and 65535), and *real numbers* (such as 0.117, 1.3, 3.14159, 17.0, 22/7 and $-4.294967295 \cdot 10^9$). The compiler is not so gifted and it needs to be instructed about the data type. It stores its (binary) version of the number in memory and needs to know how to interpret what it finds there, and how to apply the fundamental arithmetic operators $+ - *$ and $/$ for addition, subtraction, multiplication and division. The important point to appreciate is that the integer number 17 and the real number 17.000 are stored in different ways: they are represented by different bit patterns in memory, and operations performed on them must recognise this difference and deal with it. Indeed, languages such as **Fortran** and **C++** allow the programmer to extend the predefined data types and operators by defining *data structures*, or *classes* and to have control over the operators which manipulate them.

3.1 DATA TYPE: INTEGER

The basic unit in computing is the byte, consisting (usually) of 8 bits, and hence capable of storing $2^8 = 256$ different patterns. If the 8 bits are interpreted as being a binary number (i.e. the bits are numbered from 0 to 7, and a value of 2^N is assigned if bit N is set or zero if it is not) these patterns represent the numbers 0 to 255.

3.1.1 Twos-complement negative integers

But negative numbers are needed too. This is generally done by a technique known as *twos-complement representation*. The bottom 7 bits are taken as a binary number, and the top bit is assigned a value -2^7 if it is set, or zero if it is not.

So a number like 240_8, which is 10100000_2, represents $2^5 - 2^7 = 32 - 128 = -96$. The largest number is $177_8 = 127$, and the smallest is $200_8 = -128$. The number -1 is represented by 377_8, which is a byte with all its bits set.

Problem 3.1   *What is the value of   (a)* 247_8 *(b)* 324_8 *(c)* 255_{10} *if interpreted as twos-complement integers? What are the twos-complement representations of   (a)* -12 *(b)* -2 *and   (c)* -99?

Problem 3.2   *What range of numbers could be stored in a word of 10 bits using (a) unsigned integer and   (b) twos-complement representation?*

You might think this system is rather complicated and that it would be simpler to use the top bit as a sign bit

$$value = (-1)^{Bit7} \cdot (Bit0 + 2 \cdot Bit1 \cdots + 64 \cdot Bit6)$$

but the twos-complement scheme has two great advantages. One is that zero is unique. 00000000_2 is the only representation of zero, whereas any signed scheme will include $+0$ and -0 as having separate representations, so any code that tests for a value being zero (which happens quite a lot!) has to consider both possibilities. The second is that standard binary addition hardware, adding bit by bit and carrying 1 into the next column if the sum is 2 or more, works with the twos-complement system.

Problem 3.3   *Show that binary addition works for positive and twos-complement negative numbers as described, provided the result falls in the range of numbers that can be represented.*

You can find the twos-complement of any number by forming the ones-complement (i.e. replacing every 1 by a 0 and every 0 by a 1) and then adding 1 to the result. This works as the sum of the original number and the ones-complement gives the binary number in which every bit is set; adding another 1 then gives zero.

C++ distinguishes between *unsigned* and *signed* integers. A variable may be declared as **signed int** (the default) or as **unsigned int**. For a 1-byte integer the bit pattern 11111111 would be interpreted as the number -127 in the former and 255 in the latter. **Fortran** uses only signed integers.

3.1.2   Long and short integers

The integer range from -128 to 127, or 0 to 255, contains only 256 values and so integers are usually stored in more than one byte. The default storage capacity of an integer, called **int** in **C++** and **INTEGER** in **Fortran**, is decided by the compilers and the machine architecture. The default in earlier **C++** implementations is typically 2 bytes, which can store 2^{16} values. Integers with a sign hence span the values -2^{15} to $2^{15} - 1$ (-32768 to 32767). This is a remarkably small range and in practice it can be the source of apparently strange behaviour, i.e. $256^2 = 0$, as is discussed in §2.2.1.

If 2 bytes won't give a big enough range for the integer size that you're using, **C++** provides the **long** integer (at least 4 bytes) encompassing about ± 2 billion, specifically* -2147483648 to $+2147483647$. Long and short integers in **C++** are declared as **long** or **long int** and **short** or **short int**, and are **signed** by default. To use unsigned integers declare **unsigned long** etc.

In **Fortran** you can use a parameter, called **KIND**, to specify the size of values you want the variables to store. This is added to the end of the integer after an underscore. For KIND=2 you would write **1234_2**, and on our PC processor this also provides 2-byte integers with the range -32768 to 32767. In **Fortran** the default is 32 bits (4 bytes) or you could write **1234_3**, if 3 is the appropriate **KIND** value, to achieve the same effect.

Problem 3.4   *Suppose that **hi** is the largest positive integer and **lo** is the largest negative integer. Using your computer, with 2-byte integers, see whether*
1) **hi** *is 32767*   *2)*   **lo** *is* -32768

Problem 3.5   *With the same set-up as the previous problem, show what happens when you evaluate some expressions*
1)   **−hi**   *(expect to give* **lo+1**)
2)   **hi+1**   *(expect to give* **lo**. *This is a wrap-around underflow*)
3)   **−(lo−1)**   *(expect to give*   **hi**, *which is correct*)
4)   **−lo**   *(expect to give*   **lo**)

The last result shows the lower limit of the integer range is its own negative! Normally only 0 has this property. This is the consequence of finite-range integer arithmetic. The behaviour is compiler-dependent and some compilers give an error message at an out-of-range attempt. The **C++** library function **abs(int)** gives an error for **abs(−32768)** as expected with the above ranges.

Problem 3.6   *Repeat the previous problem with pencil and paper, using 3-bit integers (for which **hi** is 3 and **lo** is* -4).

---

* Actually the ANSI standard says the smallest negative **short** integer shall be at least as negative as -32767, but all the compilers we are aware of allow -32768. Likewise **signed char** and **signed long** only legally have to accommodate -127 and -2147483647, but we would be very surprised if -128 and -2147483648 failed to work.

C++ provides a total of 17 (!) kinds of integer, and 9 are detailed in Table 3.1.

Table 3.1   C++ integers (9 of the 17)

name	bytes (typical)		range	
char	1	−128	to	127
unsigned char	1	0	to	255
int	2	−32768	to	+32767
unsigned int	2	0	to	+65535
signed int	2	−32768	to	+32767
unsigned short int	2	0	to	+65535
short int	2	−32768	to	+32767
long int	4	−2147483648	to	+2147483647
unsigned long int	4	0	to	+4294967295

There are no unsigned integer representations in **Fortran** so the number of bytes used for an integer completely specifies the range. Conversely, the range of integers to be covered by a variable tells you the number of bytes you should assign to it, which is specified by the **KIND** attribute.

So far so good. But the relation between the **KIND** value and the number of bytes is not defined in the language! This is thanks to the language designers' insistence on total flexibility and future-proofing. So specifying the **KIND** of a long integer as 3 may work on our system but not on yours. To get round this, and write portable code, a function is provided – **SELECTED_INT_KIND(n)** – which gives the **KIND** parameter for integers large enough to represent n decimal digits.

```
INTEGER, PARAMETER  :: large = SELECTED_INT_KIND(9)
INTEGER, PARAMETER  :: small = SELECTED_INT_KIND(5)
INTEGER (KIND=small)  :: my_salary
INTEGER (KIND=large)  :: your_salary
```

Note that **KIND** is *not* an attribute! So it doesn't need a comma and double-colon. But it does need the brackets, to show that you're not trying to declare an integer variable named **KIND**! Having said all that, you don't in fact need to worry about the matter very often. Most integer manipulation doesn't require a huge range, and the default **Fortran** integer size (4 bytes) is usually enough. Of course you may be wasting memory by storing small integers in more bytes than they need, but unless you're dealing with enormous arrays of integers, it is a rare problem.

Example 3.1   *Sum the integers 65530 to 65540. Now sum their squares.*

Solution 3.1   *If in C++ you unwisely choose **int** or **unsigned int** (and they are 16-bit) you are back in the situation of Table 2.3. Your results will be −11 and 121 (!) while **long ints** give 720885 and −1441671, the same as **Fortran**. Whether or not you find this to be a problem depends on the calculations you are doing!*

3.1.3   Integer constants

In C++ any integer literal constant, by default, has the type **int** (and is thus signed), unless it is too big for that type in which case a larger one is used. To specify it as **unsigned,** add a 'U', or 'u' suffix, e.g. '63U'. To specify it as **long** use 'L', or the 'l' form.* Thus **1234L** specifies a long integer (signed) while **1234LU** or **1234UL** specify an unsigned long integer of value 1234, stored in 4 bytes. This is needed infrequently, as the compiler will usually convert values to the correct type in a sensible way, and the sequence

> **unsigned long int m;**
> **m = 123;**

works fine. It's not necessary to say **m = 123UL;**, though a perfectionist programmer would do so, especially when far from the declaration.

Both languages understand octal constants (base 8) and hexadecimal ones (base 16). In C++ the addition of a zero before the integer signifies that it is in octal, **023** is $2 * 8_{10} + 3 = 19_{10}$. Here is a potential source of mistakes; we are used to $016 = 16_{10}$, and not 14_{10}! Prefixing a **0x** or **0X** signifies hexadecimal, **0x23** is $2 * 16_{10} + 3 = 35_{10}$, **0xF** is 15_{10}, and **0x1AB** is $1 * 256_{10} + 10 * 16_{10} + 11_{10} = 427_{10}$.

Fortran uses the letter O for octal constants, Z for hexadecimal, and it also understands binary constants, using the letter B. The numbers are put within quotes (single or double quotes can be used): **O'123'** and **Z"1AB"** and **B'110101001'**.

3.2   DATA TYPE: FLOAT AND REAL

Variables which are used to store non-integer numbers are known as **REAL** in **Fortran** and **float** in C++. They are declared and defined in the same way as integers.

integer	age;		INTEGER	:: age
float	mass , width , height;		REAL	:: mass , width , height
float	pi = acos(−1.0);		REAL	:: pi = 3.141593

The *real numbers* of mathematics are generally written as decimal numbers, with digits before and after a decimal point, such as −273.15 or 3.14159. Such a scheme is not really suitable for numeric computation by computer (although there is a system called BCD or binary-coded decimal that does just this.) Instead, the *floating-point* system is used. This has the same idea as the notation (available on most calculators by a button marked 'ENG' or 'SCI') in which numbers are written as a value, the *significand* (also called the *mantissa*), raised to a power of 10, the *exponent*.

$$number = significand \cdot 10^{exponent}$$

This is often used for very large and very small values, such as $6.022 \cdot 10^{23}$ and

---

* But this is best avoided to minimise confusion with the digit '1'.

$1.602 \cdot 10^{-19}$, but it can be used for any number, such as $-2.7315 \cdot 10^2$ or $31.4159 \cdot 10^{-1}$. The floating-point system used by computers is similar, except that powers of 2 are easier to deal with than powers of 10.

$$number = significand \cdot 2^{exponent}$$

Various systems are in use and the most widespread by far is the IEEE system: floating-point numbers are stored in 4 bytes with one for the exponent and three for the significand. The exponent is an 8-bit unsigned number biased by 127, so a stored exponent value of 0 implies an actual exponent of -127 while the highest stored value of 255 implies an actual value of $+128$. The significand is normalised (by adjusting the exponent if necessary) so that it is always a number between 1 and 2. Then – because there are only two digits in the binary system – the leading digit is always 1 and a bit is saved by not storing it! The sign of the significand is also saved as a single bit.

The smallest number that can be stored in this representation is $2^{-127} = 5.9 \cdot 10^{-39}$. The largest is (almost) $2^{128} = 3.4 \cdot 10^{38}$. The smallest change is 1 bit in 24: $1 \div 2^{24} = 6.0 \cdot 10^{-8}$ This means that the numbers are stored with a precision in decimal of roughly 7 significant figures (see §9.1).

Problem 3.7   *The floating-point representation used in the IBM 360 series used 16 as a base for the exponent, rather than 2. Consider the advantage and disadvantage of this method.*

An important (and occasionally devastating) consequence of this is that only decimals which are also a power of 2 are exactly represented. Thus 0.625, 2.03125, 3217/1024 and -77.75265 are stored exactly but $2./3. = 0.66666667$ and 0.10 are not. The result of $(0.10 - 10.0*0.01)$ is not guaranteed to be zero when using computer reals, as it would be mathematically.

Problem 3.8   *Find (to 4 significant figures) the largest positive floating-point number your computer can handle using whatever the default floating-point representation is. Find the smallest non-zero value that can be represented. Are the magnitudes of these numbers the same if the values are negative?*

3.2.1   Higher-precision floating-point numbers

More precision than 7 significant figures, and a greater range of values than 10^{-38} to 10^{+38}, are available at the price of taking up more storage. Thus in the alternative IEEE long system, numbers are stored in 8 bytes. The significand has 53-bit precision, and the exponent (which is biased by 1023) occupies 11 bits.

This can be specified in **Fortran** by the type **DOUBLE PRECISION**, though this is really only included in the language for historical reasons: it is best to use the **KIND** parameter. The two-parameter function **SELECTED_REAL_KIND**(n , m)

returns the **KIND** value needed to store numbers with n decimal digits of precision in the range $10^{\pm m}$. (Currently, many systems use **KIND=1 , 2** for the 2 precisions.)

```
INTEGER , PARAMETER    :: dble=SELECTED_REAL_KIND(15 , 300)
REAL(KIND=dble)        :: BIG1     ! Approved way
REAL(dble)             :: BIG2     ! Approved way
DOUBLE PRECISION       :: BIG3     ! Alternative
```

The KIND of a real number is found by invoking the function **KIND(X)**

```
single  = KIND(0.0E0)
double = KIND(0.0D0)
```

In order to convert between the various possible real data types **Fortran** provides a function **REAL(A , KIND=k)** to do this (at the cost of a plurality of uses for overworked names such as KIND and REAL).

```
xdouble = REAL(X , KIND=double)
```

though such explicit conversion is not often necessary.

In C++ you have to see what your particular system provides; 8-byte reals will probably be available as the type **double**. The language definition specifies that **real**, **double** and **long double** be available as legal types, but it doesn't specify what they mean, except to say that the precision of **double** must not be worse than that of **real**, and likewise **long double** must be no shorter than **double**. Which means that although you can be sure that your program will compile on two different platforms, there's no guarantee it will give the same results! Borland's C++ v.3 for example (on the PC) gives the user 80-bit numbers with 19 places (although it works to a different arithmetic – binary-coded decimal – which is built into the 80387 coprocessor chip and its successors).

If you switch to higher-precision variables in your calculations this may not affect your program speed noticeably, or it may drive it right through the floor. A large effect results when you change a computationally intensive piece of code to (say) double precision, if the system can do operations on single-precision floating-point in hardware but has to use software library calls for the higher-precision operations.

Example 3.2   *Suppose there is a complex variable (§3.3) declared as* **COMPLEX(dble) :: zvar = (137.d0 , 1./6.d0)**. *How do you extract a single-precision representation of its real part?*

Solution 3.2   *It is rather clumsy:* **real_part_zvar = REAL(REAL(zvar))** *is needed. The first* **REAL** *does the conversion. To clarify the intention of the code,* **real_part_zvar = REAL(REAL(zvar) , KIND=single)** *is an improvement.**

---

* **single** is assumed defined as above.

3.3 DATA TYPE: COMPLEX

Complex numbers consist of a pair of reals in the order (real part , imaginary part). For example, we code $z = re^{i\theta}$ as

```
z = r*exp(complex(0.0 , theta));   |    z = r*EXP(CMPLX(0.0 , theta))
```

Everything pertinent to real-number representation also applies (twice) to complex, such as **KIND** parameters.

In **C++** implementations the complex class is not currently part of the basic language, but will be a mandatory part in the forthcoming ISO standard. It is generally provided in a header file **complex.h**. (It is intended that users create their own classes of objects, and complex number arithmetic provides a good example.) This means that you have to put **#include <complex.h>** in your program if you want to use them. The usual mathematical operators also work correctly on complex numbers, using the correct rules as in

```
complex z1 = complex(a , b) , z2 = complex(c , d) , z3 , z4;
                       // z1 = a + ib and z2 = c + id
z3 = z1 * z2;          // z3 = complex(ac−bd , bc+ad)
z4 = z1 / z2;          // z4 = complex(ac+bd , bc−ad) / (c*c+d*d)
```

The functions **real()**, **imag()**, **arg()**, **abs()**, return the obvious properties of their complex number argument, **conj()** takes the complex conjugate and **norm()** gives the square of the magnitude, **abs()**. There is also a modulus/argument form available called **polar(r , theta)**, which gives $z = re^{i\theta}$.

In the **Fortran** implementation the complex class is present as part of the basic language, but (unfortunately for the clarity of the program) it chooses to use brackets alone to indicate a complex number. We feel that this choice makes the program harder to read. However, there is a function **CMPLX(u , v)** which is legal (except in declarations) and can be used. We claim that the coding of **z1 = CMPLX(a , b)** is more readable than the allowable form **z1 = (a , b)**. The intrinsic functions **REAL**, **AIMAG**, **ABS**, and **CONJG** are available.

```
COMPLEX  ::  z1 , j = (0. , 1.)
COMPLEX(KIND(0.D0))  ::  z2
z1 = CMPLX(2.0 , 3.5)                  // z1 = 2 + 3.5 j
arg_z1 = ATAN2(AIMAG(z1) , REAL(z1))   // produces the angle of z1
z2 = j**j                              // =exp(−pi/2)=0.2078796..
PRINT * , AIMAG(z2)                    // expect 0.0 in this case
```

The purpose of using complex numbers is to simplify the programming. In many cases of physical interest the complex representation is the natural one, e.g. for FFT transformations, AC circuits, 2D rotations in the plane, and for certain continued fractions. The mathematical flow is then much clearer.

3.4   DATA TYPE: LOGICAL

Computer programs – like people – often have to make decisions

```
if ( n  <  limit )                      IF ( n  <  limit) THEN
  {
      do_something();                        CALL do_something
  }
else                                    ELSE
  {
      do_something_else();                   CALL do_something_else
  }                                     ENDIF
```

The way such choices are made is dealt with fully in chapter 4: at this point we just
mention that as part of such decision-making processes it's useful to have variables
that take on the values of TRUE or FALSE.

```
int within_bounds                       LOGICAL ::  within_bounds
within_bounds  =  (n  <  limit);        within_bounds  =  (n  <  limit)
if ( within_bounds )                    IF ( within_bounds ) THEN
  {
      do_something();                        CALL do_something
  }
else                                    ELSE
  {
      do_something_else();                   CALL do_something_else
  }                                     ENDIF
```

In **Fortran** this is done using the specific data type **LOGICAL**, as in the fragment
above. The two possible values of a logical variable are known as **.TRUE.** and
.FALSE. and you never need to know what actual bit patterns these correspond to.
In C++ there is currently no logical type: ordinary integers are used with 1 meaning
TRUE and 0 meaning FALSE. (Newer versions of the language have the type **bool**.)

In both languages, logical variables can be assigned using logical expressions,
which are formed using the relational operators 'greater than', 'less than', 'equal to',
etc. These are mostly obvious and again full details are in Chapter 4. There are a
couple of points to watch. First that the 'not equal to' operator is different in the two
languages, != in C++ and /= in **Fortran**. Secondly that the 'equal to' relational
operator is ==, *which should not be confused with the assignment operator* =. But it
always is. In **Fortran** this will probably be caught as a syntax error by the compiler.
In C++ it will probably compile, but it won't give the results you expect!

Problem 3.9   *What does* **if (k = 0) cout << "Hello World";** *actually do?
compared with the intended* **if (k == 0)** ...

In building complicated logical expressions from simpler ones care is required and
the chance of error increases. Many logical expressions are legal but nonsensical, or

not what was expected. There is a precedence, from left to right in the evaluation of logical expressions which may not be appreciated. For example, we try to write code in C++ to select a value of **n** between 10 and 20.

    valid= (10 <= n < 20)   // Does not work as it always gives 'true'

this contains two logical comparisons, **10** <= **n** which is evaluated first, and will give 0 or 1. Call it *p*. Then *p* < **20** is evaluated and it will always be true. To do the test correctly you could use the AND operator **&&** and write

    valid = ((10 <= n) && (n < 20))   // this selects n between 10 and 20

This is an example where logical variables make code more clear. Compare it with

    int large_enough = (n >= 10);
    int small_enough = (n < 20);
    valid = (large_enough && small_enough)

We would assert that the sensible use of logical variables adds greatly to readability, and is one of the signs of a competent programmer.

This also highlights the potential pitfalls of the operator precedence rules. How is the expression **a || b && c** (or **a .OR. b .AND. c** in **Fortran**) evaluated? In fact AND has a higher precedence than OR in both languages, so the expression is true if **a** is true, or if **b** and **c** are both true. It would presumably be correct to code something like **IF (credit .OR. debit .AND. within_limit)** to decide whether to allow a cash card withdrawal. But if you consider a statement that checks temperatures with **IF (monitoring .AND. too_high .OR. too_low)** there's nothing obviously wrong with it – the logical flaw doesn't (unless the programmer is very experienced) leap off the screen. Programmers deserve no sympathy whatsoever when such sloppiness catches them out. Relying entirely on rules of precedence is a bit like relying on default types. You are bound to fall into unnecessary traps of your own making.

When in doubt use parentheses! There are many situations when you as programmer know precisely what you are trying to achieve (always, we trust!) but you are unclear of every detail of the precedence rules.

Always make your intentions explicit (by using parentheses)

Another trap is the ability to mix logical and arithmetic operations in C++. In a word – don't do it! A strength of the **Fortran** logical data type is that you can't: such mixups break the syntax rules and are trapped by the compiler. If **a** and **b** have been declared as **int** in C++ and **LOGICAL** in **Fortran**), then code like **(a < 8) + b** will produce a compiler error in **Fortran**, because the addition operation can't be applied to logical variables. But in C++ it is legal: it will evaluate the logical comparison, and returns 0 or 1, which is added to **b**. Presumably this doesn't embody the programmer's (obscure) intentions.

3.5   DATA TYPE: CHARACTER

Chapter 6 is entirely devoted to characters, and this section is a brief introduction. An 8-bit byte can be used to represent a character. Generally this is done with the ASCII character set (see Appendix 2). C++ defines the type **char** which, as you would guess from the name, is used to hold these characters, though its basic definition is as an 8-bit integer (which can be signed (the default) or unsigned) and variables of this type can also be used for integer arithmetic with small numbers. **char** is the smallest size object in **C++** and other objects are integer multiples of it.

Literal constants are written* with single quotes; **'n'**, **'o'**, **'w'**, **' '**.

Some characters are unprintable, but you can write them using *escape sequences* introduced by the backslash character which is \. These are given in §6.1.4. They are of particular use when one is designing legible output formats. Most common are the newline character \n and the null character \0.

To progress from single characters to multi-character words and sentences, you use *strings*. C++ does not have a particular string type; instead it uses arrays of type **char**, with the convention that after the last character in the string there is a *null character*. C++ defines a string literal constant, with a series of characters enclosed in double quotes. For example, **"Now is the time\n"** consists of the 15 printable characters (counting the blanks), the newline character \n , *and* a terminating character (\0) which the compiler adds to signal the end of the string. These are most used in initialisation statements (and directly in output statements).

> char message[17] = "Now is the time\n";

Their use in assignments, tests for string equality etc, is tricky: see Chapter 6.

Problem 3.10   *What is the difference between* **'Q'** *and* **"Q"**? *Which of* **'a2'** *and* **"a2"** *is legal?*

In **Fortran** we find an altogether different treatment. There is a string type (called **CHARACTER**), but no 'single letter' type. Anything contained between single or double quotes is a string. So **'abcDEF 90_'** is a character string constant of length 10, **"Q"** is another, of length 1, and **"**(two single quotes) is another, of length 0. A character variable needs to be defined to be of sufficient length to hold all the characters intended for it; its length is fixed and is known to the compiler:

```
CHARACTER :: alphabet_lower*26
CHARACTER :: alphabet*52 = 'ABCDEFGHIJKLMNOPQRSTUVWXYZ'
CHARACTER :: digits*10 = '0123456789'
```

In the second case the remainder of the character variable is not initialised. It could be used, for example, to store the lower-case characters as well. These strings can be added together, called concatenation, by using the // operator, so that **'abd'** // **' cef'** would yield the string **'abd cef'** which could be stored in a variable of size at least 7.

---

* You can use **'A'** or **65** (if the ASCII set is being used) as appropriate, but it is unwise to rely on it.

★3.6   DATA TYPE: POINTER

Although *pointers* are dealt with in detail in Chapter 7, it is helpful to introduce them here. A pointer stores the address of something; e.g. a simple variable, a string, an array, or even a function. The compiler allocates variables in a defined way to memory locations and a float will (usually) occupy 4 consecutive bytes starting from its address. A pointer variable can contain this address. The operation of indexing an array is nothing but setting the pointer to the address where the array starts and then counting in units corresponding to the size of the array element type, 4 bytes for a float or a long int, 2 bytes for an unsigned int, etc. In C++ the address of a variable is obtained using the ampersand *address-of* operator, **&**, and pointer variables which can contain such values are declared using the asterisk:

```
float * f_ptr;              //  pointer to hold the address of a float variable
int *  i_ptr = &k;          //  pointer which holds the address of the int k
 /* ... */
f_ptr = &x[0];              //  f_ptr now points to element [0] of the array x[ ]
```

Knowing the address (the pointer value) we find the contents stored, beginning at that address, by using the *dereferencing operator* *,

```
q_0 = *f_ptr;               // the float q_0 contains x[0]
q_1 = *(f_ptr + 1);         // the float q_1 contains x[1]  This is the clever bit.
q_k = *(f_ptr + k);         // the float q_k contains x[k], symbolically
p_0 = *x                    // also contains x[0]
```

A pointer variable can be changed like any other, and in this manner we can perform pointer arithmetic to access the array.

```
// Print elements in array up to first zero
  for(f_ptr=&a[0]; *f_ptr != 0; f_ptr++)
        cout << *f_ptr << endl;
```

This is actually rather clever of the compiler. **f_ptr++** increments the pointer so that it points to the next object: this means adding one for a **char**, four for an **int** or **real**, and whatever is appropriate for a pointer to some other (user-defined) data type. The compiler can do this because it knows what **f_ptr** points to – a **float** in this case – and it knows how much to increment by through the value of **sizeof(float)**.

It's important to realise that in the definition **float *p**; the asterisk belongs to the pointer **p**: it is perhaps better to read it as 'The contents of **p** form a float' rather than '**p** is a pointer to float'. A multiple declaration such as **float* p1 , p2;** is amazingly deceptive: it actually declares **p1** as a pointer to float and **p2** as a float. This is almost certainly not what was intended. Get into the habit of putting in all the asterisks for clarity: **float *p1 , *p2;**. If/when you do want to declare a type and a pointer-to-type then use two separate statements.

To add readability we recommend a consistent policy on pointer names, such as beginning them all with **p_**.

3.7   USER-DEFINED TYPES

Characters and logicals can both be seen as ways of investing simple integers with particular special meanings: 65_{10} can mean A, 0 happens to mean false. Very often you may have your own particular set of meanings which you want to map onto the integers, using a number to denote a category.

For example, if you have a set of data on the height, weight, age and sex of a sample of children, you would handle the first three values as real numbers, and use an integer which you set to 1 for a boy and 2 for a girl. Or 2 for a boy and 1 for a girl. Or 0 for a boy and 15 for a girl. The scheme doesn't matter provided you stick to it consistently. You might want to code the days of the week using seven numbers, or the status of your apparatus using four (standing for Ready, Down, Transitional, and Unknown).

Writing these numbers into your program is a sure-fire recipe for disaster. Let's say you have indeed coded the day of the week in an integer variable called **today**, which is filled somewhere, and you only want to look at data taken on Mondays. Fine, you say

```
if (today  ==  1)   // ... analyse data
```

because you believe Monday is day 1. Other people start the week on Sunday, making Monday day 2. Other people (**C++/C** programmers especially) like to start counting with zero, making Monday day 0. How can you be sure that the programmer selecting the variable follows the same convention as you do? And this is a very simple and basic example!

A better way is to declare a set of constants

```
const int Mon  =  1;        INTEGER , PARAMETER  ::  Mon  =  1
    /* ... */                   ! ...
const int Sun  =  7;        INTEGER , PARAMETER  ::  Sun  =  7
```

Even if these are used by several programs/functions/subroutines in different files, there *must* be only one set of definitions. This is because one day somebody will need to change the definitions, and if they change one but not another, there will be a classic data incompatibility problem. This is handled using an include file (in **C++**) or a module (in **Fortran**) – see Chapter 5. This ensures that everyone beavering away on their different parts of the overall code is singing from the same song sheet, writing statements like

```
if (today  ==  Mon)   // ... analyse data
```

In **C++** the enumeration facility achieves this set of definitions very much more elegantly:

```
enum day{Mon , Tue , Wed , Thu , Fri , Sat , Sun};
```

sets up a type called **day** that has 7 possible values called **Mon** to **Sun**. These are coded as integers 0 to 6 but you often don't need to know that. These can be used as constants. You can declare variables to be of this type

```
day  today, tomorrow, yesterday;
```

and use them

```
today  =  Thu;
    /* ... */
if (today  ==  Fri)
{    /* ... */   }
```

In some cases the numeric code should be regarded as having no intrinsic meaning – all you can do with them is test for equality or inequality. In other cases it can be regarded as denoting something, perhaps an ordering,

```
if (today >= Mon && today <= Fri) /* ... it's a weekday */
```

perhaps an actual value, but the compiler is fussy. You can assign such a value to an **int**, but you can't assign an **int** to an **enum** type unless you use a cast

```
int i = today;              //  OK
today = 3;                  //  Illegal!
today = day(3);             //  OK
```

You can specify the values: the default for the first value in the list is zero, the default for any subsequent value is one more than the previous one. Either can be overridden if desired. It makes sense to do this if (but only if!) the values you give do have some meaning of their own rather than just being a mechanism for distinguishing different cases. Suppose you were dealing with a set of metals and you wanted to use their atomic number as the value; **metal** then contains the successive atomic numbers up to Zn = 30

```
enum metal  {Sc  =  21 , Ti , V , Cr , Mn , Fe , Co , Ni , Cu , Zn};
enum coin   {Copper  =  29,  Silver  =  47,  Gold  =  79 };
```

★ 3.8   DEFINING TYPES IN C++

What with different types, pointers, structures, storage classes and so on, a C++ variable definition can become quite complicated and hard to read. To help this you can use **typedef** to form your own shorthand. For example, if you do a lot of work with bytes and get tired of referring to them as 'unsigned char' you can insert your own definition

```
typedef unsigned char  byte;
byte  b1 , b2 , b3;              // equivalent to unsigned char b1 , b2 , b3;
```

This can be especially neat when combined with the sometimes clumsy pointer notation

```
typedef unsigned char *byte_pointer;
byte_pointer p1 , p2 , p3;              //equivalent to
unsigned char *p1 , *p2 , *p3;
```

and avoids the common trap of coding **unsigned char* p1 , p2 , p3;**

3.9 STRUCTURES

It is often useful to gather together a set of variables (perhaps of different types) that all relate to the same object. This can be done in both languages. Let's take an obvious example

```
struct element                          TYPE element
{                                          CHARACTER :: symbol*2
    char symbol[3];                        INTEGER  ::  atomic_number
    int atomic_number;                     REAL     ::  atomic_weight
    float atomic_weight;                END TYPE  element
} ;
```

which are sensible ways of describing elementary chemistry: any element has its own particular set of *components*: its one- or two-letter symbol,* its mass, the number of protons, and others can be added if need be.

Entities of this user-defined compound type can then be declared. Components are referred to using a period in C++ and a percentage sign in **Fortran**. These names can be treated as if they were any ordinary variable of the appropriate type, the only difference is that their name happens to contain a percentage sign or a period.

```
    // structure in C++                   ! structure in F90
element hydrogen , oxygen;            TYPE (element) :: hydrogen , oxygen
float water_weight;                  REAL :: water_weight
water_weight =                       water_weight =
    2.0*hydrogen.atomic_weight           2.0*hydrogen%atomic_weight
    +    oxygen.atomic_weight;            +    oxygen%atomic_weight
```

In C++ the definition of the structure and its use to declare variables can be combined into one statement, and this is often done for brevity

```
struct element {char[3] symbol...} hydrogen , oxygen;
```

or even, if you're not going to use this particular structure definition again

```
struct {char[3] symbol...} hydrogen , oxygen;
```

Initialisation is done in C++ by putting the values in sequence, separated by commas, within braces. In **Fortran** you can *construct* a constant version of a structure by using the structure name, and this can be used in an initialisation, and for other purposes.

---

* Provided we avoid the newer transuranics. The **Fortran** string always has two characters. C++ needs room for a terminator, which will be the second element for **H,O,C**... and the third for **He, Li**...

```
element carbon = {"C" , 12 , 12.0};        TYPE(element)  ::  &
                                           carbon = element('C' , 12 , 12.0)
```

Structures can be nested (i.e. their members can be other structures.) And they can contain arrays. And you can declare arrays of structures. The possibilities are endless.

```
struct pt                                  TYPE  pt
{                                            REAL  ::  x , y
   float x , y;                            END  TYPE  pt
} ;
                                           TYPE  line
struct   line                                pt  ::  from,to
{                                          END  TYPE  line
   pt from , to;
} ;                                        TYPE(line)  ::  &
                                           d =line(pt(0. , 0.) , pt(1023. , 799.))
line  d  =  {{0. , 0.} , {1023. , 799.}};
                                           REAL  ::  length
float   length_of_d  =  sqrt(              length  =  SQRT(  &
   pow(d.from.x–d.to.x , 2)                   (line%from%x–line%to%x)**2  &
   +pow(d.from.y–d.to.y , 2));              + (line%from%y–line%to%y)**2)
```

Problem 3.11   *You want to declare the 92 elements of the periodic table for use throughout your program: their atomic weights, atomic numbers, and perhaps other data too. You could do this by having a 92 element array for the weights, another for the atomic numbers, and so on as necessary. Or you could define a structure with weight and number as components, and declare an array of 92 such structures. Consider the advantages and disadvantages of the two methods. Which would you prefer to use?*

3.9.1   Object classes

Structures in **C++** can also include functions designed to operate on the data members of the structure, and they can contain private data which is not directly accessible to all of the program. A structure with this facility is known as an *object class*. Objects and their classes are important new features of the **C++** language. Classes provide for *data hiding* and *function hiding* and they provide a language extension which you can tailor to precisely whatever needs you have. The topic is discussed in Chapter 10.

3.10   ARRAYS

Arrays are collections of objects of the same data type, stored contiguously in memory. Each object is accessible by an *index*. If the data type is a real (floating-point) number then an 1D-array is essentially the same as a mathematical vector and a 2D-array is the same as a matrix. But arrays can also be defined using other data types – for example in **C++** a string is an array of type **char** – and even of user-defined data types and structures.

Arrays have a shape which, in 2D, is defined by the number of rows and columns; $A(2 , 3)$ or $A(600 , 5)$, containing respectively 6 and 3000 elements. The size of each element is determined by its type. The dimensions may exceed two, see §3.10.2.

3.10.1   One-dimensional arrays (vectors)

The individual objects are labelled by their linear position in the array. **Fortran** uses round brackets to indicate the array elements. The simplest declarations are

```
REAL ::  a(8)              ! these each declare arrays of 8 elements
REAL(KIND(0.0D0)), DIMENSION(8) ::  profile
```

The first sets up 8 locations labelled **a(1)** , **a(2)** , **a(3)** , ... , **a(8)** which are adjacent in memory.

In **C++** square brackets are used to denote an array element, which is always indexed from 0 (zero). Thus the array **c** containing 8 floats will be declared in **C++** by

```
float c[8];    //declares an array of 8 elements
```

This sets up 8 locations called **c[0]** , **c[1]** , ... , **c[7]**. Each of these can be used in exactly the same way as any other variable of that type. Please note – and this is a very common mistake – that **c[8]** is *not* one of these! Declaring an array of **n** elements means that the highest index is **n−1**.

What happens if you do stray outside the bounds and use **c[8]**, or **a(9)**, in your program? Well, the compiler will take this as being a reference to a memory location 8 full words (say) above the start of the array. But the compiler will not have reserved this location and may well be using it for some other variable. So if you read from it, goodness knows what you'll pick up, and if you write to it, goodness knows what effect it will have! An unpredictable program crash is very likely to follow. Array-bounds violation is one of the commonest sources of bugs, and also one of the hardest to track down. Although **c[8]** may stick out like a sore thumb, **c[k]** – where **k** is calculated somewhere else – looks harmless. For this reason some compilers offer an option of running programs in a more controlled way with array checking applied, which is safe but really slows the speed down.

We would argue (from our own painfully accumulated experience) that you should always check that an index is legal if there is any possibility that it may be out of range, no matter how far-fetched and unlikely this will may seem. Suppose you are going to fit a curve through a few *x* and *y* coordinates. If you say

```
int i, n;                              INTEGER :: i, n
float x[100], y[100];// store 100 values   REAL      :: x(100), y(100)
cin >> n;          // should be <=100   READ *, n   ! get number of points
for(i=0; i<n; i++)                     DO i = 1, n
    cin >> x[i] >> y[i];                   READ *, x(i), y(i) ! store points
                                       ENDDO
```

Then we absolutely guarantee that one day someone will try and use it for a data sample with more than 100 points.

Problem 3.12   *Rewrite the above fragment to trap the possibility of being given too many data points, in a user-friendly and readable way.*

Fortran is more flexible than **C++** in that other ways of indexing the locations can be specified. You can give a range of locations instead of just the size (indeed, size *n* is equivalent to specifying a range of **1** to **n**).

```
REAL :: a(1:8)          ! equivalent to a(8)
REAL :: b(0:7)          ! sets up 8 locations b(0) to b(7) like C++
REAL :: c(12:20)        ! sets up 8 locations c(12), c(13) ... c(20)
REAL :: d(-10:-2)       ! sets up 8 locations d(-10)...d(-2)
```

This can make **Fortran** programs clearer than the **C++** equivalent, e.g. when one wants an array index to range from -100 to 100 you can say so, rather than having to remember to use an offset.

```
float values[201];          REAL :: values(-100:100)
int i;                      INTEGER i
cin >> i;  // range -100 to 100   READ *, i  ! range -100 to 100
cin >> values[i+100]; // in 0 to 200   READ *, values(i)
```

As well as literal integers, constant expressions of various types can be used:

```
PARAMETER :: (N=100, J=3)
REAL :: x_value(0:N-1), y_value(0:N-1), error(0:N-1)
                       ! three 100—element arrays, (0,99)
REAL :: freq(-N/2:N/2), m_state(-J:J), VEC(0:2*J-1)   ! 3 more arrays
CHARACTER, DIMENSION(N/2) :: address*20, name*20
```

When an array is declared, the computer may set all the contents to zero. Or it may not. It may set them to 'not-a-number' (known as NaN) so that read access will be invalid till they've been properly set. It may, if it's interested in efficiency, just leave untouched whatever happens to be there already in the way of random junk. So be careful – especially if you transport a program between systems! It's usually safer to initialise them explicitly. **Fortran** provides a very general way of doing this:

```
INTEGER :: A(3)  =  (/ 11 , 13 , 12 /)          ! set to 11, 13 and 12
INTEGER :: B(3)  =  (/ 3*7 /)                    ! set to 7, 7, 7
INTEGER :: C(3)  =  (/ (I , I=1,3) /)            ! set to 1, 2, 3
INTEGER :: D(3)  =  (/ (I , I=2,8,3) /)          ! set to 2, 5, 8
INTEGER :: E(6)  =  (/ (2*I−1 , I=1,3) , 3*0 /)  ! set to 1, 3, 5, 0, 0, 0
```

To make sense of this you need just the simple facts that

- In **Fortran** you can construct an array as a list of values separated by commas and enclosed by the $(/ \quad /)$ slash–bracket combinations
- repeated values in the list can be encoded using an asterisk preceded by the repeat count and
- an *implied do-loop* can be given using a dummy variable, with range and (optional) step, all enclosed in brackets.

Example 3.3   *If an initialisation statement is*
$(/ \ (2 , 3 , J=1,4) , 4 , 3 , 2 , (J , J=8, 2, −2) \ /)$ *what does it actually give?*

Solution 3.3   *The first part expands to 2,3, 2,3, 2,3, 2,3. Then follows the set 4 , 3 , 2. The last part* $(J , J=8, 2, −2)$ *yields the value of J when it takes the values in the* **DO** *loop, J = 8 , 6 , 4 , 2. The total statement initialises the first 15 elements of a 1-D array of at least 15 elements to, e.g.*
$Quinze(15) = /2 , 3 , 2 , 3 , 2 , 3 , 2 , 3 , 4 , 3 , 2 , 8 , 6 , 4 , 2/$

In **C++** the initialisation uses braces to denote a list of elements

```
int c[3]  =  {11 , 13 , 12};       // set to 11, 13 and 12
```

but the more complicated repeat features are not available. However, there are two nice features: if the dimension is omitted then it will be taken as the number of initialisation values, so the above could have been written as

```
int c[ ]  =  {11 , 13 , 12};
```

and if the dimension is given, and is greater than the number of initialisation values, the rest of the array is initialised to zeros. So the following is good style:

```
int c[6]  =  {0};
```

3.10.2 Multidimensional arrays

Arrays may be defined in several dimensions (the number of dimensions is called the *rank* of the array). The **Fortran** declarations

$$\text{REAL} \ :: \ \text{p}(1{:}5\,,1{:}3) \ , \ \text{q}(0{:}4\,,0{:}2) \ , \ \text{r}(2\,,3\,,4\,,5\,,6\,,7)$$

declares **p** and **q** to be 2-dimensional arrays, of the same size (5·3 = 15) elements in 5 rows of 3 columns, and **r** to be a 6-dimensional array of size equal to the product of the numbers of elements, namely 5040. Notice the difference between commas (separate dimensions) and colons (give a range within a dimension) in these declarations.

A single member of an array of rank *n* is specified by *n* index integers: **p(2 , 3)** or **q(0 , j)**. The second row of **p** is **p(2 , :)** and the bottom 3 × 2 sub-array of **q** is given by **q(2:4 , 1:2)**. Much use of this is made in the array calculations in §9.8.

Fortran imposes a maximum of 7 dimensions, but this is seldom a problem in practice; 3-dimensional arrays are especially useful to hold the *xyz* coordinates for 3D transformation calculations.

It may not be obviously useful, but it can simplify programs significantly, that in **Fortran** an array can have a size 0 (see §9.8).

In **C++** the declaration uses a separate set of square brackets for each dimension. The integers are the *dimensions* (sizes) not the labels.

float p[5][2] , q[2][5] , r[2][3][4][5][6][7];

In fact the **C++** language only defines 1D arrays(!) – but the elements can be an object of any type, including arrays. So **p** above is an array of 5 arrays of 2 floats. The first element of these arrays is, in turn, **p[0][0]**, **q[0][0]**, and **r[0][0][0][0][0][0]**. The maximum element in **p** is **p[4][1]**, while in **q** it is **q[1][4]**. The only things that **C++** allows us to know about the array are a pointer, specifying the address of the first element, and the type of the elements. Accessing the array elements is nothing but pointer manipulation. Pointers and arrays are very, very closely linked in **C++**, as will be discussed in Chapter 7.

The ordering in memory of the elements of a multidimensional array is different in the two languages! In **Fortran** *the first index runs fastest* i.e. the array is in column-order. The 6 elements of **ftn(2 , 3)**, with its 2 rows and 3 columns, are stored in memory as **ftn(1 , 1)**, **ftn(2 , 1)**, then **ftn(1 , 2)**, **ftn(2 , 2)**, then **ftn(1 , 3)**, **ftn(2 , 3)**. In **C++** *the first index runs slowest*, i.e. the array is in row-order. The 6 elements of **cpp[2][3]**, with its 2 rows and 3 columns, are stored in memory as **cpp[0][0]** , **cpp[0][1]** , **cpp[0][2]**, then **cpp[1][0]** , **cpp[1][1]** , **cpp[1][2]**. Once you get above the most basic level in dealing with arrays, these differences matter. The storage determines the order of printing when the whole array is accessed.

For instance, in **Fortran** an array may be moulded into another shape by the use of the function **RESHAPE**: to create a 3-by-2 array from the 6 element rank-1 array x you use:

y = RESHAPE (SOURCE = x , SHAPE = (/ 3 , 2 /))

Problem 3.13   *If you do the above, is* **y(1 , 2)** *set to* **x(2)** *or* **x(4)**? *Show that the processing is by column and that the second possibility is correct.*

If you're using large arrays you may want to worry about memory caching: the helpful way that a CPU may transfer chunks of data from slow memory to fast onboard memory (or from disc to RAM in 'virtual memory' systems). If you're using an array so large that the whole thing is not cached, then a loop which repeatedly accesses elements from members of the array that are far apart in memory can cause the processor to spend a lot of time caching and de-caching data. If you can confine your computations to elements that are close together the caching overhead is minimised. Such fine tuning needs care and may produce surprises.

Problem 3.14   *You are designing a subroutine in* **Fortran** *or a function in* **C++** *as part of a gas-simulation program. The gas will consist of* 100000 *identical particles characterised by their positions x, y, z and their velocities* $\dot{x}$, $\dot{y}$, $\dot{z}$ *all to be stored in the array* **gas()** *or* **gas[ ]**. *Each loop through the calculation will want to access all the* 6 *properties efficiently for each of the* 100000 *particles. Decide whether your array should be dimensioned as* 100000 *rows and* 6 *columns or as* 6 *rows and* 100000 *columns.*

3.10.3   Array operations in Fortran

Fortran – which is generally superior to **C++** when it comes to array operations – has an elegant feature whereby an entire array may be referred to by the name with corresponding operations being performed on each element in turn, individually. Thus if **x(6)** and **x_square(6)** have been declared then

    x_square = x**2

is equivalent to

    DO k=1 , 6
        x_square(k) = x(k)**2
    ENDDO

and similarly for

    x = SQRT(x_square)   ; z = EXP(x) ; x = x − 2.0 ; z = x / y
    x = 5.5   ! equivalent to x(1)=5.5; x(2)=5.5; ... ; x(6)=5.5

and so on. When the RHS is a constant then all elements of the array are set to the value of the constant.

This is a welcome feature. **Fortran** simplifies the programmer's task with these constructions and it will take advantage of any parallel processing hardware available.

3.10.4   Array sections in Fortran

Arrays can be moulded, manipulated and subdivided in **Fortran**. By the use of sections any definable subarray can be referenced. As an example suppose we have a 4 by 4 array **A(4 , 4)** as the normal matrix of fitting data to a cubic equation (§9.8). The lower right 3 × 3 subarray can be written as **A(2:4 , 2:4)**, and can be used in array operations: for example if we want to copy it to the array **SUB** we can just put **REAL SUB(3 , 3)**= A(2:4 , 2:4). Likewise the whole of column 2 can be referred to by **A(1:4 , 2)** or **A(: , 2)**.

The section, e.g. **2:4** can also be given as a *vector subscript*, say **INTEGER :: u(3)** = (/2 , 3 , 4/) and the reference made by **SUB** = A(u , u). The vector u can be created more generally as u = (/i : k : j/) meaning the implied **DO** loop **DO i , k , j** from **i** to **k** in steps of **j** (which can be positive or negative). The only restriction is that the entries in **u** must be present in the matrix/array it is applied to.

Suppose we have a 4000-channel data vector stored in **peaks(4000)** containing features of interest, such as absorption peaks from a laser scan or X-ray peaks in a silicon detector, and that each occupies 40 channels or so. Then an elegant way to process each peak is to select the appropriate range for u = (/L : U/) and work on **peaks(u)**, by plotting the data, finding the backgrounds, centroids, areas, etc. By varying **L , U** you can march up and down the data spectrum as you choose.

The vector subscript is more general and can contain its values (the subscripts) in any order and with repetition allowed.

There are also useful masking operations which allow operations to all elements which satisfy a certain criterion. The construct **WHERE** is used. A simple example is

```
REAL :: DEE(6)  =  (/ −7. , −5. , −3. , −1. , 1. , 3. /)
WHERE ( DEE > 0.0 ) DEE = SQRT( DEE )
```

which produces (−7.0 , −5.0 , −3.0 , −1.0 , 1.0 , 1.732) because negative elements failed the logic test and $\sqrt{1.0} = 1.0$, $\sqrt{3.0} = 1.732$. Such operations are possible, of course, in **C++** but they must be constructed:

```
float dee[6]  =  { −7. , −5. , −3. , −1. , 1. , 3. };
for( int i=0; i<sizeof(dee); i++)
    if ( dee[i] > 0.0 )
        dee[i]  =  sqrt(dee[i]);
    /* ... */
```

4

Control

Two roads diverged in a wood, and I —
I took the one less traveled by,
And that has made all the difference.
– Robert Frost: 'The Road not Taken'

Logical expressions are first introduced, and the variables, constants and operators that can be used in them. Then follows the use of such expressions in the various forms of conditional statement, beginning with branches and ending with loops. After discussions of different approaches, an example is given to show how the same process can be programmed using several different techniques.

Real programs contain sections of code which, depending on circumstances, need to be performed once, or several times, or not at all. This chapter explains the various ways by which this control of execution is achieved using *conditional statements*.

In both languages the syntax for a simple condition is sensible and obvious.

    if (condition)   statement; | IF (condition)   statement

For example

    if (bal > 0.0) cout << "In credit\n"; | IF (bal < 0.0) PRINT * , ' Overdrawn'

The *condition* in round brackets is a logical expression, which is evaluated. The result, which is either true or false, determines whether or not the following statement is executed.

While you can do a lot with this simple choice, more sophisticated forms are available for programming convenience. The story continues in §4.2, but first, logical expressions are dealt with in §4.1.

4.1   LOGICAL EXPRESSIONS

In both languages the **if** statement *evaluates a logical expression* to determine whether or not the specified action is to be taken. Because there are two and only two possible outcomes (i.e. the action either is taken or it isn't) the expression can only have two possible values, called TRUE and FALSE. Although they are most often met with as part of a conditional statement, they can also appear in other contexts (such as assignments to logical variables) and their use can lead to more compact, readable and elegant programs.

These expressions are often quite simple, for example a relational operator between two arithmetic expressions, or they can be severely complicated, also including *logical operators* acting on *logical variables* and other logical expressions.

A logical variable, also known as **boolean** after its inventor George Boole, can store the results of logical expressions. The *logical operators* are AND, OR, XOR and NOT – language specific details are in §4.1.1 and 4.1.2. The expression (A AND B) is TRUE if and only if A and B are both TRUE. The expression (A OR B) is TRUE if either A or B, or both, is TRUE. The OR operator is the *inclusive* OR, whereas the XOR operator is the *exclusive* OR (also called EOR) and is only TRUE if A is the opposite of B. The truth table for logical operators is a summary of these relations:

Table 4.1   The truth table for logical operators

A	B	A AND B	A OR B	A XOR B	NOT A
FALSE	FALSE	FALSE	FALSE	FALSE	TRUE
FALSE	TRUE	FALSE	TRUE	TRUE	TRUE
TRUE	FALSE	FALSE	TRUE	TRUE	FALSE
TRUE	TRUE	TRUE	TRUE	FALSE	FALSE

A complicated logical expression may involve several different operators. Their precedence is complicated (§2.4.6), but one rule is useful enough to be committed to memory: in both languages the arithmetic operators come before the logical ones, so that x + 3 > 2 * y means (x + 3) > (2 * y). It may also be helpful to remember that NOT comes before AND and AND comes before OR, so

workday = weekday AND NOT holiday OR overtime

does what it should, whereas

picnic = (saturday OR sunday) AND NOT raining

needs the brackets.

You can't remember all the rules. (Well, we can't.) You can look them up in tables of precedence (e.g. Tables 2.2 and 2.3). If in doubt use brackets. In fact we suggest that even if you are in *no* doubt, put them in. It will make the code easier to read.

4.1.1   The Fortran type LOGICAL

In **Fortran** there is a specific type **LOGICAL**. Variables of this type can be declared just as any other. There are also *logical constants*, but (in contrast to the enormous number of possible integer and real constants), here there are only two: **.TRUE.** and **.FALSE.** – where the periods are part of the syntax. The separate logical type gives **Fortran** the advantage that you can't, either on purpose or by accident, perform inappropriate operations. Having declared **LOGICAL :: a , b** then expressions like **a+b** or **3*b** will get flagged as errors by the compiler.

For example one could use a logical variable **file_open** in a program to flag whether some file was open or not, that is required for reading or writing.

```
LOGICAL :: file_open = .FALSE.
OPEN(1 , FILE='results.dat')          ! open the file on channel  1
file_open = .TRUE.                    ! I/O status = .TRUE.
   ! ...                              ! read/write using the file
IF (file_open) THEN
   CLOSE(1)                           ! close the open file on channel  1
   file_open = .FALSE.
ENDIF
```

The use of logical variables is a sign of an effective programmer. You could have accomplished the same effect using **file_open** as an integer variable, setting it to 1, say, if the file is open and 2 if it is shut. But this is arbitrary, is less clear to read and is thus liable to lead to mistakes. A logical called **DEBUG** (or something similarly informative) is often a useful tool in program development. You declare it and set it to **.TRUE.**, and at suitable places in your code you output potentially useful information if the value is set.

```
LOGICAL :: DEBUG=.TRUE.
   ! ...
DO i = 1 , n_data
   fit_value = slope*x(i) + const
   IF(DEBUG) PRINT * , i , x(i) , fit_value
      ! ...
ENDDO
```

When your program is working smoothly, you just change the **.TRUE.** to **.FALSE.** and the information is suppressed. Then when you hit a case where it stops working properly and you need to investigate, you just set it back to **.TRUE.**

The *logical operators* in **Fortran** are **.AND.**, **.OR.** and **.NOT.**: there is no exclusive-**OR** operator.

Problem 4.1   *How would you evaluate the exclusive* **OR** *of* **A** *and* **B**?

How does **Fortran** store its logical values? It isn't easy to find out and indeed you need never know the actual bit patterns your **Fortran** compiler uses for TRUE and FALSE values. This may vary between different compilers.

4.1.2   Logicals in C++

C++ today has no special logical type (unlike **Fortran**). Instead it uses ordinary integers (type **int**, or even **char**); 0 is taken as corresponding to FALSE, and anything else (usually 1) as TRUE.* Again there are three operators: "!" for negation, "&&" for AND, and "||" for OR. So you can put together expressions like

month_valid = (month > 0) && (month <= 12);

Although these are 'logical operators' there is strictly no difference in C++ between logical, relational, and arithmetic operators since (**i** && **j**) and (**i** == **j**) are expressions just like (**i** + **j**). Contrast this with **Fortran** where **i** .AND. **j** is only valid if **i** and **j** are of type logical, in which case **i** + **j** is not legal.

There is one further complication. The values 0 and 1 represent FALSE and TRUE, but what about other numbers, 2, 3, 4 ...? Only zero (FALSE) is unique so when a condition is evaluated and interpreted in a control statement, the action hinges on whether the value is zero or non-zero. Any non-zero value thus has the same effect as the value one, and the condition is treated as TRUE.

Problem 4.2   *Why is the statement* **x_logic** = **!!x_logic**; *useful? Compare it with the closely related expression* (**x_logic** ? **1** : **0**) *(see §4.4.1).*

The proposed introduction of the new type **bool** with literal values **true** and **false** will bring the advantages of the **Fortran LOGICAL** type to C++ programming. Their values seamlessly merge with 1 and 0 in any comparisons. If your compiler doesn't provide them then you may construct your own pseudo-logical constants with

const int true = 1;
const int false = 0;

This can be good technique in the constant struggle for clearer code.

But there are some uses for the older system: for example some function **fit_data** could return an integer which is an error code: zero if all is well, and various non-zero values for particular disasters. This can then be tested using the fact that *all* the non-zero values will be flagged as true.

```
if (i = fitdata())
{                         //... take action to handle error
}
else
{                         //...carry on processing
}
```

Problem 4.3   *At first glance you might think there should be a double equals sign in that* **if** *statement. What is actually going on?*

---

* This is just a convention – other languages, like PL/I, do the opposite. Yet others, such as BBC BASIC, use –1 for TRUE and 0 for FALSE. This conveniently made the bit patterns all ones or all zeros.

4.1.3   Relational and logical operators

Logical expressions can involve not only the three logical operators AND, OR, and NOT, but also the *relational operators*. These are

<	>	<=	>=	==	!=	in C++
<	>	<=	>=	==	/=	in **Fortran** and also
.LT.	.GT.	.LE.	.GE.	.EQ.	.NE.	are equivalent **Fortran** forms*

The meaning of these is clear. Notice that 'greater than or equal to' and 'less than or equal to' are >= and <= (with the = sign last), are in English language order. The 'equals' operator is == to distinguish it from = (the assignment operator). Treating the two languages together allows us to remind ourselves of the very different meanings of the /= symbols. In **Fortran** it is the logical operator 'not equal' whereas in C++ it means 'divide by'. Thus a **Fortran** eye would have considerable trouble with the statement **a = b /= c;**.

These operators can be used in expressions, giving a logical result.

pow(b , 2) > 4.0*a*c;	b**2 > 4.0*a*c
month <= 12;	month <= 12
i + j + k == 17;	i + j + k == 17

All evaluate to 1 or 0 in C++ and to **.TRUE.** or **.FALSE.** in **Fortran**, and can be used in conditional statements, or assigned to (logical) variables, or used in more complicated logical expressions.

The legality of these relational operators depends on the data types they're being used with, and follows common sense. For example, you can't use > between two complex numbers in **Fortran**. But beware; **x == 1.0** will evaluate as FALSE if **x** is 0.999999 or 1.00001, and arithmetic rounding will produce unpleasant surprises for you. Try evaluating expressions like $(1 == 1/3 + 2/3)$ or $(2.0/3.0 == 1.0/3.0 + 1.0/3.0)$ and see how far you get (§9.1 considers this further). Sometimes you may get away with it, but it is always bad practice and should be completely avoided.

Logical expressions involving strings will be explained in detail in Chapter 6, for now we just note that in **Fortran** they are straightforward, whereas in C++ they can be a diabolical trap. If **name** has been declared as a character string, then expressions like **name == "Query"** and even **name >= "aardvark"** are legal in both languages. In **Fortran** they do what you would expect, but not in C++! Basically, C++ applies the comparisons between the *addresses* of the strings, rather than to the strings themselves. Special-purpose functions are available for string comparison (§6.3.10).

Both languages deal with comparisons between variables of different types in the most sensible way they can manage. In carelessly programmed comparisons like **if (x > i)**... the real number **x** will be compared with the integer **i** by converting the value of **i** to a real number. In some cases such a conversion is impossible (you can't compare a number and a string) and then the compiler will flag it as an error.

---

* Their interest is now purely historical. It's a shame that **Fortran** keeps the clumsy .**AND**., .**OR**. and .**NOT**. formats when it managed to replace these other dinosaurs.

In **Fortran** you can't use $==$ and $/=$ between *logical* values. Instead (and why this is so is not at all clear) the operators **.EQV.** and **.NEQV.** must be used, so to determine whether the ordering of **a** and **b** is the same as that between **c** and **d** you use: **filter = (a > b) .EQV. (c > d)**.

The repetition of the symbols in **C++** for the AND (**&&**) and OR (**||**) logical operators is to distinguish them from the bitwise AND (**&**) and bitwise OR (**|**) operators which work on *each bit individually* (§2.4.5). The bitwise version of NOT (**!**) is the inversion operator $\sim$. Confusion is possible and care is needed.

Problem 4.4   *If you have declared* int **legal=6** , **honest=3;**, *what are the values of* **(legal & honest)** *and* **(legal && honest)**?

The problem may look artificial, but it brings out the fact that both forms are legal and perfectly sensible, and it's very easy to code one instead of the other.

4.1.4   De Morgan's laws

Complicated logical expressions can often be written in many ways, some of which are more readable, or more efficient, than others. In such cases it is helpful to use de Morgan's laws which relate to *logical negation*. Suppose there is an integer **n** and you wish to test whether it has the value 2, 3 or 4. Then you may write either of

IF (n==2 .OR. n==3 .OR. n==4) ! ...| IF (n > 1 .AND. n < 5) ! ...

To do something if the condition is *not* fulfilled we must have, in the second case, (N is not greater than 1) OR (N is not less than 5). In symbols, for both:

IF (.NOT.(n==2) .AND. .NOT.& | **IF (.NOT.(n > 1) .OR. .NOT.(n < 5))! ...**
(n==3) .AND. .NOT.(n==4))! ... |

These spell out *de Morgan's laws of logical negation*: where **L** and **M** are logical variables, then

$$\text{NOT (L OR   M) = (NOT L   AND   NOT M)}$$
$$\text{NOT (L AND M) = (NOT L   OR   NOT M)}$$

i.e. you negate the separate parts, and change the intervening operators from AND to OR, and vice versa (taking extra care with precedence brackets to avoid unexpected errors). It may well be clearer to create logical variables with mnemonic names, and combine them. Contrast the readability of the above examples with:

LOGICAL :: GreaterThan_1 , LessThan_5
GreaterThan_1 = (N > 1) ;   LessThan_5 = (N < 5)
IF (GreaterThan_1 .AND. LessThan_5) THEN ! ...

Problem 4.5   *I can walk to work or take the bus. I take the bus if it's raining or if I'm late, and provided I have enough money for the fare. When do I walk? (Apply de Morgan's rules.)*

4.2   CONDITIONAL STATEMENTS

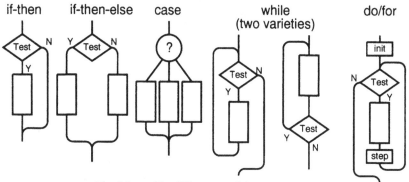

Fig. 4.1   The different conditional statements

There are five basic patterns of conditional statement, given schematically above:

1 **Simple condition**: a block of code is executed if required, but not otherwise. If a bank balance is negative then a warning message is printed. This is done by the simple **if** statement.

2 **Alternative condition**: one of two blocks of code is executed. A program to solve quadratic equations $ax^2 + bx + c = 0$ will need to take different actions depending on whether $b^2 - 4ac$ is positive or negative.* This is done by using the compound **if—then—else** statement.

3 **Multiway condition**: one of several blocks of code is executed. For example a menu program will perform one of many possible actions, according to the selection made by the user. This is done using the **case** construction.

4 **Iteration**: a block of code is repeated until some condition is met. (The condition may be tested before or after the block.) For example a program to solve nonlinear equations by Newton's method will repeat the process several times until it gets the desired accuracy. Use **while, for, DO** loops for this.

5 **Indexed iteration**: a block of code is repeated a number of times. For example a routine to calculate the scalar product of 3-dimensional vectors will multiply appropriate pairs of components together three times. This is done using **for** loops in **C++**, and the analogous **DO** loops in **Fortran**.

These five types are not strictly different, but the distinctions are useful in practice. These situations are commonly met with, and language tools are provided to deal with them. Using an inappropriate tool – for example a string of **if else** statements instead of a **case** – is possible, but it's generally like undoing a nut with a pair of pliers because you're too lazy to find the right spanner.

---

* The zero option (equal roots) can be treated as positive provided that (as in **C++**) **sqrt(0.0)** does not return an error.

4.3   THE SIMPLE CONDITION: IF   STATEMENT

The simple and straightforward   **if** (**condition**)   **statement** was introduced at the start of this chapter. The round brackets are mandatory. For example

if (k > N) cout << "No more data\n"; | IF (test < 1E−6 .AND. iter > 100) z = 0.0

The *condition*, in brackets, is an expression which is evaluated. The result, which is either TRUE or FALSE, determines whether or not the following statement is executed. If there is a whole group of statements whose execution depends on this test, this is done similarly:

```
if ( x < 0.0)                        IF (x < 0.0) THEN
{                                      ! This is the block
     cout  <<  " x is negative \n";    PRINT * , ' x is negative = ' , x
     y = 0.0;                          y = 0.0
}                                    ENDIF
```

For **Fortran** this second form is an addition to the language (technically, it is the **IF** *construct* as opposed to the **IF** *statement*) with the keywords **THEN** and **ENDIF** that define the block of statements to be executed. For **C++** the two forms are actually the same: the executed statement can be simple, as in the first example, or a compound statement, as in the second. A statement in **C++** needs a final semi-colon even if this comes before a final curly bracket (PASCAL programmers please note!). If the block is very long then it may be better to put it into a separate subroutine/function and revert to a simple **if** statement, as this may make the code easier to read and understand.

In statements like this it is important that the structure is immediately apparent to anyone looking at it. Indentation is vital, and the use of spaces helps readability. Contrast these two:

```
IF(x>0.0) THEN                       IF ( x > 0.0 ) THEN
a=b                                    a = b
PRINT *,' value of c is ',c            PRINT * , ' value of c is ' , c
ENDIF                                  ENDIF
PRINT *,' e= ',e,' f= ',f              PRINT * , ' e = ' , e , ' f = ' , f
```

The **IF** and **ENDIF** statements have to be seen to match. With **C++** you have to make it clear how the braces (curly brackets) match, to enclose blocks of code. We think that the clearest program format is the one above, i.e. (i) Each brace is on a line of its own (ii) braces are indented (opening and closing braces having the same amount of indentation) and (iii) the contents of the block are indented further. Contrast these,

```
if ( x < 0.0 )                       if ( x < 0.0 ){
     {cout << " x is negative \n";     cout << " x is negative \n";
         y = 0.0;}                      y = 0.0;}
```

as examples of two less clear versions, with that given above.

4.4   THE COMPOUND **IF—ELSE** STATEMENT

If your program has to choose between two alternatives, the one to be executed if the test condition is true, the other if it is false, then in both languages this involves an extended form of the **if** statement using the keyword **else**:

```
if (dev < tol)
    cout << "converged\n";
else
    cout << "no convergence\n";
```

```
IF (dev < tol) THEN  ! condition TRUE
    PRINT * , ' converged'
ELSE                 ! condition FALSE
    PRINT * , ' no convergence'
ENDIF
```

If you have an **ELSE** in **Fortran** it has to be on a line of its own, between a **THEN** and an **ENDIF**.

More complicated situations may require **if** and **if—else** statements to be *nested*, i.e. the blocks of code may contain one or more further **if** statements:

```
if (balance < 0.0)
{
    cout << "Overdraft\n";
    if ( !arrangement )
    {
        cout << "Not permitted!\n";
    }
}
```

```
IF (balance < 0.0) THEN
                ! outer condition
    PRINT * , ' Overdraft'
    IF( .NOT. arrangement ) THEN
                ! inner condition
        PRINT * , ' Not permitted!'
    ENDIF
ENDIF
```

Fortran provides the form **ELSE IF—THEN** . These two programs are equivalent:

```
IF(condition1) THEN
    block1
ELSE IF (condition2) THEN
    block2
ENDIF
```

```
IF(condition1) THEN
    block1
ELSE
    IF (condition2) THEN
        block2
    ENDIF
ENDIF
```

But notice the difference in the **ENDIF** statements. When **THEN** occurs in an **ELSE—IF** statement then it does not require a matching **ENDIF** statement of its own.

C++ does the same with the **else if** ... form. (However, this is not an extra part of the language: the second block just happens to be a conditional statement.)

You can test for a whole string of conditions using one **else if** or **ELSE IF** after another:

if (condition1)
{
    block1
}
else if (condition2)
{
    block2
}
else
{
    final block
}
	IF (condition1) THEN
	    block1
	
	ELSE IF (condition2) THEN
	    block2
	ELSE IF (condition3) THEN
	    block3
	ELSE
	    final block
	ENDIF

This code executes the appropriate block for the conditions, testing in order until one is satisfied – if none of them is met, then the final block is executed.

It is vital to ensure that the conditions do not overlap. If **condition1** is **k < 10** and **condition2** is **k > 8** then **k = 9** will only be dealt with by the first condition, despite what the programmer may have had in mind.

A real-life example is that of a temperature controller when a decision must be made according to the current temperature, stored in **Temp_now**. Within a (±) temperature bandwidth about the desired reference temperature **T_ref** some action is required. The logic may be organised as follows:

```
IF    (Temp_now > T_ref+Bandwidth) THEN      ! Temperature too high to control
          allow_to_cool
ELSE IF (Temp_now > T_ref) THEN              ! start control
          apply_integral_control
ELSE IF (Temp_now > T_ref—Bandwidth) THEN    ! full control actions
          integral_and_proportional_control
ELSE                                         ! T too low so apply max heat
          heat_fully
ENDIF
```

You can build up quite complicated structures in this way. This is usually a mistake. Many such decision trees are better handled by the **case** construct, to be mentioned shortly.

In fact the nested **if** statement is responsible for some of the most unreadable code around. Many programs involve several **if** statements that are checked before taking action. For example, when you read out your apparatus to store a reading, you may want to check first that it is switched on and working, and that the data is there and is reasonably sensible. You build up an edifice something like

if (volts_on)
  { //...ask for readout condition
    if (readout_working)
    { //...try and take data
      if (data_present)
      {//...check data not rubbish
        if (data_valid)
        { //... analyse data
	IF (volts_on) THEN
	  !...ask for readout condition
	    IF (readout_working) THEN
	    !...try and take data
	      IF (data_present) THEN
	      !...check data OK
	        IF (data_valid) THEN
	        !...analyse data

This is all clear and nicely indented with it. The nightmare comes at the end:

}
      }
      else
      {
        cout << "error!\n";
      }
    }
  }
	ENDIF
	
	ELSE
	
	      PRINT * , 'error!'
	      ENDIF
	    ENDIF
	  ENDIF

It is totally unobvious which condition gives the error message. And this is an artificially clean and simple example: in a real program all this will be spread over many lines with lots of intervening material.

Fortran provides a feature to help. An **IF** statement can be *named*, and this name given attached to the corresponding **ENDIF** and **ELSE** statements:

```
VOLTAGE:     IF (volts_on) THEN
READOUT:        IF (readout_working) THEN
PRESENT:           IF (data_present) THEN
VALID:                IF (data_valid) THEN
                         ! ... analyse data
                      ENDIF VALID
                   ELSE PRESENT
                      PRINT * , 'Error: no data present'
                   ENDIF PRESENT
                ENDIF READOUT
             ENDIF VOLTAGE
```

This is not just decoration. The compiler will check that your nested **IF**s are in a consistent sequence, thereby removing one abundant source of programmer error.

In **C++** all you can do is use civilised common sense: add comments and make messages meaningful, or think about rewriting the program to use some sensible subprogram structure.

If all the conditions are intended to select the valid reading of the data with no branching actions to be taken, then the logic may be combined into a compound test:

```
ok_to_read=(volts_on && readout_working && data_present && data_valid);
if (ok_to_read) read_data();
```

The evaluation stops the first time a decision can be reached. If this were at condition 2 then conditions 3 and 4 would not even be tested. But the position of the first false condition is not available: if the test fails then you don't know where. A better solution, with a clearer structure, is

```
if (volts_on && readout_operational)
{
        if (!data_present)
                cout << "error! no data present\n";
        else if (data_valid)
                read_data();
}
```

4.4.1 The conditional operator in C++

Very often the only action taken in an **if—else—** sequence is to assign a value to a variable. As in, for example, using **Fortran**,

```
IF (itry == 0) THEN
        tol = 999.0
ELSE
        tol = 3.0*epsilon
ENDIF
```

It may be tempting to rewrite this as

```
tol = 999. ;  IF (itry /= 0) tol = 3.0*epsilon
```

which is only two statements long instead of five, and is probably no slower to execute. However it's offensive to some programmers, as the first assignment is executed even under conditions when you know it will immediately be overwritten.

While **Fortran** programmers must wrestle with the aesthetics of this, the **C++** designers spotted the need and provided a neat way out: the *conditional operator*. This uses the query and colon characters in the syntax:

$$test\text{-}condition\ ?\ value\text{-}if\text{-}test\text{-}true\ :\ value\text{-}if\text{-}test\text{-}false$$

thus the above can be written

```
tol = (itry==0 ? 999.0 : 3.0*dev);
```

The brackets are not strictly necessary, but are generally advisable as the operator precedence of **?** is very low.

4.5   MULTIWAY CHOICES: CASE/SWITCH

When the CPU signals to the output device that it wants to transmit data
it must be prepared for one of four possible replies
1: Device ready – send data
2: Device busy – do not send data
3: Anything else
4: Nothing at all
– Apocryphal

Suppose a variable is expected to have a discrete range of possible values, which specify one of several possible actions to be taken. Nested **IF** structures become unwieldy: **CASE** (**Fortran**) and **switch** (**C++**) are the civilised alternative. For example, suppose your program contains a variable called **option** obtained from the user in response to the message: **Type A to analyse this data, S to save your results, Q or E to quit.** Then you could organise the logic with

```
switch(option)
{
      case 'A':
            analyse();
            break;
      case 'S':
            store();
            break;
      case 'Q'

      case 'E':
            exit();
      default :
            cout  <<  "Invalid option!\n"
}
```

```
SELECT CASE(option)

      CASE('A')
            CALL analyse

      CASE('S')
            CALL store

      CASE('Q')
            STOP ' Q entered'
      CASE('E')
            STOP ' E entered'
      CASE DEFAULT
            PRINT * , 'Invalid option!'
END SELECT
```

The apparent similarity conceals a basic difference. In **Fortran** the desired block of code is executed, and control then passes to the statement after **END SELECT**. In C++ control is merely transferred to the appropriate point: the statement is executed, and *program flow continues to the next one and not to the end of the block*. That's why the **break** statements (§4.8) are needed: they take the program control to the end of the block – i.e. the end of the switch statement (so Figure 4.1 is not strictly accurate). In the 'Q' case this C++ feature is used: control 'drops through' to the next statement. You don't need a **break** after **exit()** because the program stops anyway.

In **Fortran CASE** statements – which can also be named for readability, like **IF** statements – you can specify a range of values, but this is not provided in C++ .

```
CASE('1':'9')
      PRINT * , ' Give a letter, not a number'
```

4.6   INDEFINITE ITERATION AND WHILE LOOPS

The simple **if** statement tests a condition and, if true, executes a statement, or block of statements. One obvious use for this is to take remedial action if a condition is unsatisfactory.

if (error > tolerance)
{
    error = refit(data);
}
	IF (error > tolerance) THEN
	
	    error = refit(data)
	ENDIF

But suppose this remedy doesn't work first time? The **WHILE** statement is designed for this. Think of it as an **IF** statement which is applied repeatedly until the condition is false.

while (size > storage)
    getmore(storage);
	DO WHILE (size > storage)
	    CALL getmore(storage)
	ENDDO

The test is only applied once every iteration, at the start. The whole block stands or falls together. Something is expected to happen partway through the block which can change the condition to FALSE but it is only when the test is re-evaluated at the start of the next iteration that the program realises this and transfers execution to the next statement after the block.

There is a school of thought that objects to the **while** statement on the grounds of machine efficiency, arguing that you can use a **for−** or **DO−**loop and a **break** (or **EXIT**) command to the same effect. See Metcalf and Reid* (1996a, §12.3.2). However the **while** version is in our view more readable and can display a different mental attitude to iteration.

4.6.1   Forever loops

In some cases you want a loop to repeat 'forever': for example, you might want a simulation program to run indefinitely, or at least until someone explicitly stops it. **Fortran** has a specific construction to do this − you just omit the **WHILE** − but **C++** does not. The **for(; ;)** construct in §4.7 provides one way of doing this, another is a **while** loop with a condition which is always TRUE. (1) is the simplest way of doing this, as by definition (1) is TRUE.

while (1)
{
    get(option);
    act(option);
}
	DO
	    ! Fortran forever loop
	    CALL get(option)
	    CALL act(option)
	ENDDO

---

* Who comment on potential problems in a parallel-processing environment.

If **while(1)**... offends you, you may prefer to use expressions such as (1 == 1) which are clearly TRUE whatever convention is in use, or if your compiler supports the **bool** type then you can use the clearer and more transparent **while(true)**. Even if it doesn't, you can do this if you have declared **true** as a **const int** with the value 1, as recommended in §4.1.2.

In practice such loops are terminated by some exceptional condition which leads to the execution of an **exit()** or **STOP** to kill the program, or of a **break** or **EXIT** to quit the loop.

4.6.2   Testing at the end of the loop: do—while

In the **C++** construction **do.. while**

```
do
{
    error = refit(data);
} while (error > tolerance);
```

the only difference is that the the condition is tested at the *end* of each iteration, rather than the beginning. This only matters as far as the first iteration is concerned, but it also means that the block is executed at least once, no matter what the condition is.

This may be a good or a bad thing. In the above instance, suppose **error** has been evaluated by an earlier part of the program and is in fact very small. Then the **do—while** form would clearly be inappropriate because you don't actually need the improvement. But if **error** has not been calculated at all at this stage, then the **do** form ensures that it is evaluated and set up before the test, whereas the **while** form would start (disastrously) by testing an undefined quantity.

This facility is not provided in standard **Fortran** which could be thought a deficiency in the language.

The comma expression (see §2.4.8) is particularly useful in a **while** statement. When two expressions are separated by a comma, a single expression is formed, whose value is the value of the second expression *after the first has been evaluated.* The value of the first expression is thrown away. For example

```
while (cin >> i , i >= 0) perform(i);
```

which reads a number from the input, and does something with it, *until a negative number is entered.*

Problem 4.6   *Contrast this with* **while (cin >> i && i >= 0) perform(i);**. *Does it work? Is there any difference? Which would you use?*

4.7 INDEXED ITERATION

Here we go round the mulberry bush,
the mulberry bush, the mulberry bush...
– Children's rhyme

In a loop (**DO** loop in **Fortran** and **for** loop in C++) a statement, or set of statements, is repeated a given number of times, using an *index variable* whose value changes by some regular step with each iteration. Although this can be done using a **while** construct, indexed loops are used so frequently that provision of a special facility to do this is justified.

Although the **Fortran** and **C++** loop constructs do pretty well the same job, they look very different. The **Fortran** syntax is

DO *index-variable* **=** *initial value, limit-value, step-value*

    ...

ENDDO

where the final *step-value* defaults to the integer 1 if it is not given. Note carefully that the loop variable must not be a real but must be an integer. This represents a change from **Fortran 77**.

The **C++** syntax is

for (*initial-statement* ; *test-expression* ; *iteration-expression*)   *statement*

where *statement* can be a simple statement terminated by a semi-colon, or a compound statement enclosed in braces.

The **C++** version is much more flexible: conversely the **Fortran** version decides a lot more for you. Contrast the following two program fragments, which do the same thing. In both cases the integer **k** is the index variable.

`for (k=1; k<=12; k++)`	`DO k=1 , 12`
`    cout << k << " men went to mow\n";`	`    PRINT * , k , ' men went to mow'`
	`ENDDO`

In each case there are four features of the loop. They are explicitly shown in the **C++** version, and implicitly in the **Fortran** one.

1 **Initialisation.**
In **C++** the initial statement, the first in the bracket, is performed. In **Fortran** the value of the index variable is assigned to the first value after the = sign. The divergence of function is already shown here. In **Fortran** that's all you can do – initialise a single variable. In **C++** a statement is a statement. It could contain several parts:

`for (int k=1, int j=2;  ... )  // declares and initialises two temporary variables`

2 **Testing.**
In **C++** the test expression (the second in the bracket) is now evaluated: if it is FALSE (zero) then the loop terminates, without the loop body being executed at all, and control passes to the next statement after the **for** loop, but if it is TRUE (non-zero) the process continues. In **Fortran** the index variable is compared with the limit, which is the second value specified on the **DO statement**. If the variable exceeds the limit (or, when the step size is negative, if the limit exceeds the variable) then the loop terminates* and control passes to the next statement after **ENDDO**.

3 **Execution.**
The body of the loop is executed. This is the block of code between **DO** and **ENDDO** in **Fortran**, or the statement (simple or compound) after the final round bracket in **C++**.

4 **Incrementation.**
In **Fortran**, the step-value is added to the index variable. The increment often has the value 1: other values can be specified as a third number in the **DO** statement itself, e.g. **DO ichabod = 10, 100, 5** or **DO infill = 10, 1, −1**. In **C++** the third expression in the bracket is evaluated. This usually involves incrementing the index variable, although **C++** enables you to do other things too, usually with the aid of the comma operator.

Problem 4.7   *What is the purpose of making the first item in brackets in the **C++** form a statement rather than an expression?*

The **C++** **for** loop can do everything the **Fortran** **DO** loop can, but much, much more. The three expressions are usually an initialisation, a test of range, and an increment, but they don't have to be – and they can do more. When it is said that 'You can write **Fortran** programs in any language' one of the things this refers to is the unimaginative use of **for** statements according to the **DO**-loop pattern. Code full of statements such as **for (i=1; i<=n; i++)** or, even worse, using **for (i=1; i<=n; i=i+1)** is a sign of a **Fortran**-trained programmer who is trying to write in **C++**, but has still not explored very far in the new language. Indeed R*E*A*L hackers like to put so much into the **for** loop control that nothing further is required except a semicolon to mark a null statement. The following **C++** fragment adds 100 elements of an array, from **a[0]** to **a[99]**,

```
for (sum=0.0 , i=0;  i<100;  sum += a[i++])   ;
```

(Note the semicolon closing the empty statement.) Much sounder code (robust to changes of array sizes and types) is

```
size_a = sizeof(a) / sizeof(a[0])
for (sum=0.0 , i=0;  i<size_a;  sum += a[i++])   ;
```

---

* **Fortran IV** programmers please note that the test – step 2 – is applied *before* the first time the loop is executed. If this first test fails then the loop is not executed at all.

In **C++** it is often useful to use the comma operator to form multiple statements, in any of the three expressions, so that several changes can proceed together: you can reverse in vector **b[ ]** the contents of vector **a[ ]** with

```
for (k=0 , m=99; k<100; k++ , m--)   b[m]  =  a[k];
```

or even,

```
for (k=0 , m=99;  b[m] = a[k] , k<99;  k++ , m--)    ;
```

The increment block does not have to be in the third expression, it can be placed within the loop.

```
for (k=0 , m=99; k<100;  )   b[m--]  =  a[k++];
```

though we find both these versions harder to read than the first. If the index variable, **k** in this case, is altered in the body of the loop, then it makes for really unclear code. In **Fortran**, altering the index in the loop is forbidden and a reputable compiler will refuse to let you do it. We rather think that a reliable programmer should also refuse to do it *in any language*.

One rather subtle difference between the two languages can be seen if (in a misguided spirit of mischief) you alter the conditions within the body of the loop.

```
n = 10;
for (k=1; k<=n; k++)
{
    n = 20;
    cout << k << " men went to mow\n";
}
```

```
n = 10
DO k=1 , n
    n = 20
    PRINT * , k ,' men went to mow'
ENDDO
```

The **C++** loop will run 20 times, whereas the **Fortran** version will run 10 times. Why? Because in **Fortran** the test value is evaluated once and for all at the loop initialisation, whereas in **C++** the second expression is evaluated every time, so that the limit could be dynamically changed (usually a very risky practice).

Loops can be nested – and in **Fortran** they can be named to make the code clearer.

```
for (i=0; i<n; i++)
{
    for (j=0; j<n; j++)
    {
        c[i][j] = 0.0;
        for (k=0; k<n; k++)
            c[i][j] += a[i][k] * b[k][j];

    }
}
```

```
ROW: DO i=1 , n

COLUMN: DO j=1 , n
            c(i , j) = 0.0
SUM:        DO k=1 , n
                c(i , j) = c(i , j) + &
                    a(i , k) * b(k , j)
            ENDDO SUM
        ENDDO COLUMN
    ENDDO ROW
```

Finally, what about the value of the loop variable when the loop is ended? In **Fortran** it is not defined. It is helpful to regard the variable as a transient whose scope lies within the loop. If you want to know the index value when some condition has forced the loop to end then copy its value inside the loop to a variable defined for the purpose. Don't think you know how the compiler works. The result may even depend on other processes which are happening in the computer; the reason has to do with the flexibility given to **Fortran** to choose which registers to use. (Putting an index variable in a register rather than memory may well help the program run faster, as it will be used a lot, for array indexing as well as the iteration test.) In **C++** the index variable has no special status, so it can be used according to normal rules. The value is the last value before the loop was exited, unless it was declared in the initialisation part of the **for()** statement, in which case it is inaccessible outside the block.

★ 4.8   ESCAPING

With a single bound he was free!
– Anon

There are some instances where you realise partway through the block of code that you don't want to go on. Both languages provide (different) ways of breaking out of the standard flow of control. **CYCLE** in **Fortran** and **continue** in **C++** take you right out of the current iteration and on to the next one. Their more drastic counterparts **EXIT** and **break** take you out of this iteration and out of the whole loop, moving on to the next program statement, whatever the state of the loop index.

Note: this is a point where the two languages use different names for similar purposes. Don't get confused: **CONTINUE** also exists in **Fortran**, for historical continuity, but is an obsolete, equivalent form of **ENDDO**. In **C++**, **exit** is a function causing *program* termination rather than *loop* termination as it does in **Fortran**.

If loops are nested, then the naming facility in **Fortran** enables **EXIT** and **CYCLE** to specify which loop control you want to interrupt. Otherwise (i.e. in **C++**, and in **Fortran** if names are not specified) the loop broken out of is the innermost one at the time.

Problem 4.8   *Code a program to solve an equation, such as $x - cos(x) = 0$, using Newton's method using 10 iterations, or until the agreement between successive answers is better than 0.0001, whichever is the sooner. Functions* **f(x)** *and* **df(x)** *should be provided to calculate the function and its differential. Newton's method finds a new estimate x_{n+1} of the root from the previous one x_n by the expression $x_{n+1} = x_n - f(x_n)/f'(x_n)$.*

★ 4.9   EXCEPTION AND ERROR HANDLING

There are times when your code can handle all the cases it considers, and times when it can't. If your processing logic is many layers deep, it can be messy to get out of: it's as if you are finding your way through a maze of twisty little passages, and discover that what you seek is impossible, and you just want to get out.

C++ has a way of doing this, called **try** − **throw** − **catch**. The code in which problems may arise is enclosed in a *try block*: the keyword **try** followed by the maze of logic, included in the usual braces. If, within this, things become too desperate to handle, the code can **throw** control away. This switches control to the *catch block* which comes straight after the try block. This takes an argument (typically a character string) which is passed to it by **throw**. If nothing is thrown, then nothing is caught, and the catch block just gets bypassed.

```
for(int i=0; i<100; i++)
{
    try
    {
        cin >> x;
        if ( x < 0.0 ) throw "Negative parameter input\n";
        f(x , i);
    }
    catch (char* message) // trap any exception in last block
    {
        cout << "Crash because " << message << endl;
        exit(1);
    }
}
    /* ... */
void f(float x , int i)
{
    if ( x < 0.000001 ) throw "parameter too small\n";
        /* ... */
}
```

The power of this mechanism lies in the way that exceptions can be thrown from inside functions – and you can't do that with a **goto** statement! (see §4.10.)

There is no such facility on standard **Fortran**, though some compilers provide it.

When you are writing serious software that handles real-time data, you can't ignore million-to-one chances that lead to division by zero, negative roots, and similar errors. Computations must be approached in a proper spirit of paranoia. When you're going to read values of **a** , **b** , **c** and compute $a/(b-c)$ every 10 milliseconds for a month then you're liable to crash the program with a division by zero, even if this logically shouldn't happen, and even if it doesn't happen on your test run of 100 values.

★4.10   THE GOTO STATEMENT

The **goto** statement is a relic of the past. It certainly arises naturally in low-level languages whose elements correspond closely to the actual operation of the CPU. But it forces a very fixed thought-pattern on the concept of program control, it leads to the writing of impenetrable code, and its place (if any) in higher-level languages is very debatable.

Nevertheless both these languages contain it, and it can be used to change program control to another statement, labelled in **Fortran** by a number and in **C++** by a legal name, elsewhere in the same routine.

```
if ( x >= 0.0 ) goto positive;                    IF( xx >= 0.0 ) GOTO 10
        cout  <<  "Negative x\n";                          PRINT *,' Negative xx'
        x = −x;                                            xx = −xx
positive:   y = sqrt(x);                           10     zeno = SQRT(xx)
        /* ... */                                          ! ...
```

In both cases this is poor and clumsy code and the following is better:

```
if ( x >= 0.0 )                                    IF ( xx >= 0.0 )
        cout  <<  "Negative x\n";                          PRINT *,' Negative xx'
y = sqrt(fabs(x));                                 zeno = SQRT(ABS(xx))
        /* ... */                                          ! ...
```

The **goto** is clearly a very powerful and general statement, and it was *the* work-horse of early programming. Yet any serious worker in computing will tell you that the **goto** statement is a BAD THING. Why is this?

Goto statements make programs hard to read and understand, and hence harder to maintain and more prone to the introduction of bugs. In a program full of labels and **goto**s the flow of control, indeed the flow of thought, is obscured. It may seem that this is a minor point of fastidious aesthetics – who cares what a program looks like provided it works? – but the defects are so severe, and the effort required to rewrite and debug incomprehensible code so significant, that it really is true that good programmers today just do not use them.

The objections of the purists culminated in E. W. Dijkstra's classic paper 'Goto statement considered harmful' (1968) which pointed out that if you look at a program listing, containing **if–then–else** statements, and are told that the computer is at a particular statement, then you can deduce the complete state of the program – which variables have been set, and which may not have been. **Do loop** type constructs make things only a little worse, since for each loop you have to know the value of the index variable. But if **goto** statements are involved, you can't do this, as you don't know the history by which the program reached the statement. Control can jump backwards and forwards. Look at the following versions of a **Fortran** subroutine (both contain a bug).

```
SUBROUTINE quod(a , b , c)            SUBROUTINE quad(a , b , c)
REAL  ::  a , b , c , x1 , x2 , test , disc   REAL :: a , b , c , x1 , x2 , test , disc
test = b*b − 4.0*a*c                 test = b*b − 4.0*a*c
IF (test < 0.0) GOTO 10              IF (test < 0.0) THEN
disc = SQRT(test)                       PRINT * , ' Complex Roots'
20  x1 =−(b + SIGN(disc,b))/(2.0*a)   ELSE
x2 =   c/x1                              disc = SQRT(test)
PRINT * , ' Roots' , x1 , x2         ENDIF
RETURN                               x1 =−(b + SIGN(disc,b))/(2.0*a)
10  PRINT * , ' Complex Roots'        x2 =   c/x1
GOTO 20                              PRINT * , ' Roots', x1, x2
END  SUBROUTINE quod                 RETURN
                                     END  SUBROUTINE quad
```

Problem 4.9   *What is the problem? In which program is it easier to spot?*

There is arguably* a case for **goto** statements where the flow is clearly forwards, and where it avoids deeply nested blocks. The example in §4.4 could also be coded as

```
if (!volts_on) go to nbg;            IF (.NOT.  volts_on) GOTO 999

    //... ask for readout condition       ! ... ask for readout condition
if (!readout_working) goto nbg;      IF (.NOT.readout_working) GOTO 999

    //... try and take data               ! ... try and take data
if (!data_present)                   IF (.NOT.  data_present) THEN
{
    cout  <<  "Data not present\n";       PRINT * , ' Data not present'
    goto nbg;                            GOTO 999
}                                    ENDIF
    //... check data is not rubbish       ! ... check data is not rubbish
if (!data_valid) goto nbg;           IF (.NOT.  data_valid) GOTO 999

    //... analyse data                    ! ... analyse data
                                     RETURN        ! All done
return(0);        // All done        999 PRINT * , ' Error occurred'
nbg:   cout  <<  "Error\n"; return(1);   END SUBROUTINE
```

They can also be used − carefully − to break out of deeply nested loops. This is particularly helpful in C++, where **break** will only break out of the innermost level, whereas the **Fortran** equivalent **EXIT** can break out to any level using named loops.

---

* i.e. We argue about it.

★ 4.11   OTHER WAYS OF THINKING

The linear if–then–else approach is only one way way to analyse control problems. People don't think that way all the time – and indeed, even computers don't. Other perspectives are often very fruitful.

★ 4.11.1   State models

One alternative approach to if–then–else analysis of §4.4 is the *state model*. Let's take that old cliché from so many traditional IT courses: the flowchart that describes how to make a cup of tea. This is clearly codable as a sequence of IF statements IF (kettle_empty) CALL fill(kettle) and so on. Having done that, it's left as an exercise to the student to draw up a flowchart for making a cup of coffee.

Now suppose that you and your friend want a cup of tea *and* a cup of coffee. The combined algorithm cannot be simply obtained from the two separate ones – if you follow your simple flowcharts, you're going to end up boiling two kettles.

A state-model approach would start by defining the *objects* that comprise your universe (e.g. kettle, teapot, and cup) and then the *states* they may be in (e.g. kettle–empty, kettle–full–cold, kettle–full–hot...). The state of your universe is defined by the states of all these objects. Next you define the *rules* by which objects change state (e.g. the fill–kettle operator takes kettle–empty to kettle–full–cold . . .). Then we have the problem solved in the form that we know that the state cup–of–tea is reached from the state (cup–with–milk, teapot–full) by the operation pour–teapot; cup–with–milk is obtained from (cup–clean, milkjug–full*), and so on. Further complications like coffee and hot chocolate can be implemented by adding further objects, states, and rules, without having to rewrite any of the existing analysis.

```
SUBROUTINE make_clean_cup(cup_state)
INTEGER      ::  cup_state
INTEGER      ::  cup_full , cup_dirty , cup_clean
PARAMETER(cup_full=1 , cup_dirty=2 , cup_clean=3)
DO
    SELECT CASE(cup_state)
        CASE(cup_clean)
            RETURN
        CASE(cup_dirty)
            CALL wash_cup(cup_state)
        CASE(cup_full)
            CALL empty_cup(cup_state)
    END SELECT
END DO
END SUBROUTINE make_clean_cup
```

---

* Or, rather, milkjug–notempty.

The contrast between this approach and the conventional flowchart is that it is free from history. States are described in a way which does not depend on how they got there. The state **kettle—empty** may arise because the last user left it that way, or because we've filled 10 cups and need a refill or because it's just been bought from the shop. The state model does not discriminate between them, whereas in a flowchart this condition would arise at several points, and the fill operation have to be incorporated separately at each.

Some languages, such as PROLOG, are specifically suited to this approach. They can be used to build *state engines* which determine the best path from your present state to the desired final state. C++ and **Fortran** are not designed in this way, but the ideas can still be used in problem analysis, and can be implemented using case statements as in the program. Here **wash_cup** and **empty_cup** are primitives: **wash_cup** will presumably change **cup_state** from dirty to clean, whereas **empty_cup** changes it to from a full cup of cold tea to an empty but dirty one. The parameter statement and declaration should be included in all subroutines using some preprocessor, or set up in a module; this is the only point at which the link between the states and the integers that represent them is made. Notice the forever loop: processing continues until the desired state is achieved. Error messages can be handled by **cup—runneth—over**.

Many adventure games are basically state models: you can't kill the dragon without the sword, you can't open the door to the armoury without the key ... and so on.

★ 4.11.2   Concurrency

Truly concurrent systems, in which an area of memory is shared by two or more CPUs, are scarce. But in an interrupt-driven multi-tasking system, where the operating system shares out CPU time between many different processes, these processes are effectively concurrent. The problem also arises in *distributed systems* where several CPUs are connected to one another by a network.

Care has to be taken when these processes communicate. They may need to ensure that two processes don't both try to use hardware which requires *exclusive access* – two programs must not send characters to the printer simultaneously – or to ensure that a task performed by one process is complete before another process uses the results: the *producer–consumer* problem.

If the processes concerned fail to communicate, then nonsense will generally result. On the other side, one has to be sure that situations cannot arise in which two over-cautious processes *lock out* and wait for each other forever, or even the extreme case of *deadlock* in which all processes are waiting for another to complete and the computer hangs up in what is termed a *race condition*.

The difficulty and interest of such problems are that in a simple sequence like **if(i_printer_flag == 0) i_printer_flag += 1;** there is no guarantee that the result of the second statement will set **i_printer_flag** to 1, as between the inspection in the first line and the assignment in the second, some other process may *interleave* with its own instructions which set **i_printer_flag** to its own value.

With simple special-purpose tasks processes can communicate by setting flags in memory: a consumer process sets a flag to zero and then starts the producer process, which sets the flag to 1 to show it is complete; then the consumer can safely wait until this happens. For less simple situations this gets tricky: if the producer is not started by the consumer but running independently, there may be interleaving problems in the setting of the flag.

Problem 4.10   *If your machine code has instructions which increment and decrement a memory location (such as* **INC** *and* **DEC** *on the 6502), show how this could be used for a flag system for the mutual-exclusion problem.*

A common solution to this problem is by means of *semaphores*. These are not part of the standard of either language, but may come as part of an extra library of real-time routines. A semaphore is a variable which is 1 (or > 1) if a resource is available and 0 if it is not. It is used in two routines usually called **lock(s)** and **unlock(s)**, where s is the semaphore variable. So in a particular problem, a semaphore is declared (and initialised to 1): when a process wants to use it, it calls **lock**, which inspects the semaphore, and if it is non-zero it decreases it by one and execution continues; if it is zero then the process is suspended. When it has finished it frees the resource by calling **unlock**, which starts a suspended process if there is one, if not it increases the semaphore by one. These two routines run in non-interruptable mode, so that they cannot be interleaved (don't ask how – that's the responsibility of the expert who wrote them). The system is straightforward and easy to use. But if a process misbehaves – for example by issuing a **lock**, and then crashing before it calls **unlock** – then this can lead to lock-out and deadlock.

★ 4.11.3   Event-driven code

For those familiar with the traditional programming format this method requires a degree of inside-out thinking. You have to drop the idea of a program, with a beginning and a middle and an end, and instead focus on the concept of an *event*.

An event is a signal to the computer – i.e. to some central overseeing program, perhaps the operating system – that something has happened. It could be something from external hardware (a temperature sensor reports a reading above the safety level), or a user (someone clicks a mouse button) or from within the program, *signalled* by other pieces of code. You write routines that *handle* each particular type of event. Your code then consists of a preamble, in which you establish the logical connection between the event types and your individual *call-back routines*, and the routines themselves. After the preamble has been executed, your program then awaits events and, when it finds one, calls the specified routine. This clearly has an ancestry in the writing of interrupt-handling routines. Visual Basic (on the PC) is a nice example of an event-driven language. Various graphic objects (buttons, slide-bars...) are defined. The program then provides routines which define what to do if various actions (clicking, dragging, entering data...) are performed on these objects.

4.12   AN EXAMPLE IN CONTROL

> *There's more than one way to skin a cat.*
> *– Traditional*

To see how the same end can be achieved by different constructions, with differing degrees of effectiveness and elegance, let's write a code fragment that asks the user for a number, and checks that the answer is sensible.

(a) using a **do–while** construct

```
do                                        i = −999
{                                         DO WHILE ( i<1 .OR. i>10 )
   cout << "Choose number, 1 to 10\n";       PRINT * , ' Type a number, 1 to 10'
   cin >> i;                                  READ * , i
} while (i < 1 || i > 10);                 ENDDO
```

Notice the clumsy initialisation of **i** to an out-of-range value needed to make sure that the **Fortran** loop is entered.

(b) with a **while** and a **break** or **EXIT**

```
while (1)                                 DO
{        // infinite loop                     ! infinite loop
   cout << "Give a number, 1 to 10\n";        PRINT * , ' Enter a number, 1 to 10'
   cin >> i;                                  READ * , i
   if ( i>0 && i<=10 )                        IF ( i>=0 .AND. i<=10 ) EXIT
      break;                                  ! failure condition
   cout << "Must be 1 to 10!";                PRINT * , ' Between 1 and 10!'
}                                         ENDDO
```

This has an additional helpful message when the user needs it.

(c) Our choice (but see §8.3 and §8.4)

```
                                          PRINT * , ' Guess a number, 1 to 10'
cout << "Select a number, 1 to 10\n";     READ * , i
while(cin >> i , ( i<1 || i>10) )         DO WHILE( i<1 .OR. i>10 )
{                                            PRINT * , ' Need 1 to 10! Try again!'
   cout << "From 1 to 10! Try again!\n";     READ * , i
}                                         ENDDO
```

There are other ways too, using **goto** statements for example. Consider how all these achieve the same (or similar) ends in different ways, and how some are more elegant and user-friendly than others. Computing is not like mathematics, where a problem has (usually) only one right answer: just because you know a way to achieve something in a program, it does not follow that it is the only way, or the best way. However old a dog you may feel, always be prepared to learn new tricks!

5

Subprograms: Functions and Subroutines

It is explained why programs are split up into subprograms: how these are coded and how they are accessed. The arguments used to pass information to (and from) subprograms are considered in detail. Reasons for building libraries of functions are given.

Typical adults can keep about 7 different things in their head at once. Any more than that and they lose track. In daily life we solve this problem by grouping of concepts together: when we plan our day the tasks of consuming coffee, cereal, toast and marmalade can all be conveniently grouped under the heading 'breakfast'; only when we've got to the kitchen do we need to consider the constituent parts.

Likewise if your program is going to do more than half a dozen different things, to keep track of them all, and the connections and dependencies between them, you have to group tasks together; then at each stage of program development you're not considering too many different activities. So the ability to split up the code into different parts is a vital aspect of all computer languages. Each part has its own purpose, which can be considered, written, and sometimes even tested separately from the rest of the program. Reading-in-data can be isolated from plotting-a-graph, or saving-results, or calculating-an-integral, and yet all are part of the same overall program. This reduces large clumsy programs to small, manageable parts, each dealt with by a separate *subprogram* (sometimes called a *procedure*). This chapter describes how this is done.

5.1   SUBPROGRAMS: FUNCTIONS AND SUBROUTINES

Suppose you are analysing a set of 100 data measurements and you want to know the arithmetic mean of these values. You decide to store the values in an array. Getting the data is one task. Extracting the mean is another. So your program could look something like this:

```
#include <iostream.h>
// function prototypes
void get_data(float data[ ] , int n);
float mean(float data[ ] , int n);

// main program
main()
{
  float data[100];
  get_data(data , 100);
  cout << "Average " << mean(data,100);
}

void get_data(float d[ ] , int n)
  {
    for (int i=0;  i<n;  i++)
      cin >> d[i];    // get data
  }

float mean(float a[ ] , int m)
  {
    float sum=0.0;
    for(int i=0;  i<m;  i++)
      sum += a[i];
    return sum / float(m);
  }
```

```
PROGRAM analysis
  REAL , DIMENSION(100) :: data
  REAL , EXTERNAL :: mean
  CALL get_data(data , 100)
  PRINT * , ' Average ', mean(data , 100)
  STOP ' normal exit : analysis '
END PROGRAM analysis

SUBROUTINE get_data(dat , n)
  INTEGER :: j , n
  REAL , DIMENSION(:) :: dat
  DO j = 1 , n
    READ * , dat(j)
  ENDDO
  RETURN
END SUBROUTINE get_data

REAL FUNCTION mean(a , m)
  INTEGER :: m
  REAL , DIMENSION(:) :: a
  REAL :: sum=0.0
  DO j = 1 , m
    sum = sum + a(j)
  ENDDO
  mean = sum / REAL(m)
  RETURN
END FUNCTION mean
```

Each main program calls two subprograms, one to read the data (probably from a file rather than the keyboard), the other to calculate the mean. The two subprograms are short and sweet and their purpose should be clear: details of the syntax are given in the rest of this chapter.

Fortran distinguishes two varieties of subprograms: *functions*, like **mean** in the above example, which return a value, and *subroutines*, like **get_data**, which don't. In C++ there is only one variety: the function (but a C++ function need not return a value and, even if it does, it may be ignored.) Indeed, C++ programs are collections of functions, one of which is called **main**. The economical C++ language uses one language element (the function) instead of three (function, subroutine, and program).

5.1.1   The subprogram body

Subprograms are independent program units, which communicate through argument lists, a return value (if applicable), and possibly also global data. Internal details are invisible to the invoking program. Most of the statements are the sort you can meet with in a main program, but a few are specific to subprograms.

5.1.1.1   Specifying that this is a subprogram

The **Fortran** statement **SUBROUTINE get_data(dat , n)** informs the compiler that what follows, up until the matching **END SUBROUTINE get_data** statement, is the body of code comprising the subroutine to be known as **get_data**. The statement also tells the compiler that this subroutine has two arguments, which will be known in the code that follows as **dat** (the data vector) and **n** (its length). The types of these, and arguments internal to the subroutine, must be declared in the subroutine. If the subroutine has no arguments then the parentheses are omitted.

The syntax for functions is almost identical, with **FUNCTION** instead of **SUBROUTINE**. A function must always have at least one argument so that the list, and the parentheses, are mandatory.* The type of the return value may be declared with the syntax **REAL FUNCTION mean(dat , n)** (as is done here) or within the subprogram, using **REAL :: mean**.

In C++ there is no function keyword analogous to **FUNCTION**. The word 'function' does not appear in the code: you just give the name of the function, preceded by the type of its return value. If there is no return value, then you use the keyword **void**. After the name comes the list of arguments. If there are no arguments, the parentheses must still be given, with nothing between them (or the keyword **void**). The body of the function follows as a statement – almost always a compound statement in curly brackets – after the argument list. Notice – it's a common coding mistake – that there is no terminating semi-colon after the final curly bracket.

In addition to defining the function C++ needs to supply the *prototype* (§5.2.1) so that the compiler can make sense of references to the function before it meets its full definition. The lines at the start of the program

```
void get_ data(float data[ ] , int n);
float mean(float data[ ] , int n);
```

are the prototypes for the functions defined later; the square brackets specify that an array is being given, without specifying its length.

Round brackets (parentheses) are a characteristic signature of a function in C++. They surround the argument list of the function. Their appearance specifies that this is a function call, and the function to be called is the name preceding the parentheses – unless this is in turn preceded by a type name, which shows that the function is not being invoked but defined.

---

* In 99% of cases it's a sensible rule that any function producing a value must have an argument. The other 1% are cases like random number generators, or calls to the time of the system clock. These generally end up with dummy arguments to satisfy the syntax.

5.1.1.2   The return

For a **Fortran** subroutine, control is passed back to the calling program by the **RETURN** statement. Actually this also happens if execution gets to the **END** statement, but we regard this as poor style. Unimaginative programmers* regard the subroutine as a single block: you work through it and **RETURN** when you get to the end. But **RETURN** is much more flexible than that, and can usefully appear as an action in **CASE** statements, or **IF (...) RETURN**. (On the other hand some stylists insist that there be one, and only one, way to exit from a subprogram.)

When control passes back to the calling program from a **Fortran** function, through an explicit **RETURN** statement or by coming to the end, then it returns as a value whatever was (last) assigned to it in the subprogram. In the above example, the line

mean = sum / **REAL**(m)

allocates the value to the function **mean** – note that it appears here *without* any argument list – and the function then passes control back to the calling program.†

In C++ **return** followed by an expression will return the value of the expression. If the function type is declared as void you can use **return;** on its own, or you could use the fact that control returns if you reach the end of the function body.

main is a function: in most instances nothing is returned. (Strictly it should be coded as **void main**, as if no return type is given the default is **int**, not **void**, but many compilers do tolerate **main** on its own.) It is sometimes useful to end it by **return 0;** if it completes successfully and **return 1;** (or some other non-zero value) if something goes wrong. (Giving the type with **int main()** helps legibility if you do this.) This value is then available to the operating system that called your program, and this can be used by scripts that run several programs in sequence.

5.1.2   Invocation

Subprograms can be called by the main program, and by other subprograms.

To invoke a **Fortran** subroutine the keyword **CALL** is used. Control passes to subroutine. Eventually the subroutine finishes its appointed task and returns control to the program that called it; control continues at the statement after the **CALL**.

Functions, in both languages, are not called explicitly, and they are just used in an expression. The system knows that they will provide a value needed to evaluate the expression, and so it passes control to them when it needs this value. Again, when the function code has found the desired value it is returned and the evaluation of the expression continues.

That makes good sense in the context of expressions like

---

* The list of **Fortran** 'obsolete features' includes something called the 'Alternate Return'. Certain of the above-mentioned unimaginative programmers have taken this to mean that in future, when all these features are rightly removed from some future **Fortran**, there will only be one **RETURN** statement allowed in each subroutine. Not so! The 'Alternate Return' was a cumbersome feature which allowed a routine to return to different places in the calling routine; it was a return to alternate labels and nothing to do with alternative **RETURN** statements.

† That means that you could use the function value as storage for a partly completed calculation, though this may be inefficient and is not a good idea e.g. if speed matters.

$$q \; = \; \log(y) \; + \; 13.4 * \text{mean}(\text{data}, n)$$

which invokes the system function **log** as well as your own function **mean**. It still applies, less obviously, to **C++** function calls that look more like subroutine calls.

```
get_data(x, 20);
printf(" Value for unit %i is %f"; 99 , sqrt(2.0F));
```

Here the expression is evaluated: to do that a function **get_data** or **printf** is called. This value – which for **get_data()** is void anyway – is then just thrown away.

Problem 5.1   *Find what value is actually returned by* **printf** *(see §8.1.1)*.

★ 5.1.3   The stack

It does help to know a bit about how the calling mechanism works. Although it's not legally specified by the language descriptions, what generally happens is that subprogram calling is done by the use of the *stack*.

A stack is a particular way of storing data. It comprises a dedicated area of memory and a pointer variable: the *stack pointer*. Initially this points to the start of the dedicated area. To *push* data onto the stack you store it at the contents of the stack pointer, which is then incremented. To *pop* data from the stack you decrement the stack pointer and read the data that it's then pointing to. In practice there are additional details like checking that the stack hasn't overflowed, but the general idea is like a stack of papers on a desk: you can add and retrieve documents, but only on the top.

Such a system is flexible and easy to use. It's so general that some hardware systems have a special register (SP for Stack Pointer) with specific hardware instructions for pushing and popping. The drawbacks are that you have to allocate the memory area beforehand, and if you're not sufficiently generous the stack may grow till it fills all available space, whereupon the program will crash. There must also be an iron discipline about the use of the stack. If a program puts a value at the top of the stack and then passes control to another procedure, this second procedure is free to push and pop items onto the stack, but it's vital that when control passes back to the first program the stack is in the state it expects to find it.

So a typical set of actions performed by the CPU when a subprogram is called (details may differ) is

- Push the current contents of the PC, incremented by 1, on the stack. This is the *return address*.
- Push each argument onto the stack.
- Jump to the start of the subprogram.
- Pop the arguments from the stack into appropriate temporary variables.
- Execute the code of the subprogram.
- Pop the return address from the stack.
- If there is a return value, push it onto the stack.
- Put the return address in the PC.
- If there is a return value, pop it from the stack.

5.2 ARGUMENTS

The *arguments* are used to pass information from the calling program to the function or subroutine, and sometimes vice versa. This is an important method of *sharing data*, §5.3.4. Others are global variables, §2.1.3, modules, §5.2.2, and internal subprograms, §5.3.3. These differ in the degree of data hiding achieved.

The arguments to be passed are given in the code that calls the subprogram, and are then available to that routine when called. Their particular values will be different from one call to the next. In the subprogram code an argument has to be given a name. In the calling program it may have the same name, or a different name, or it may be an expression that doesn't have a name at all. And if the subprogram is invoked at several different points, it may be invoked with different arguments:

```
#include <iostream.h>
  void out(float q);
  main()
  {
      float x=3.14159F , y=2.71828F;
      out(x); out(y); out(17.3F);
      out(x*x+y−0.3F);
  }
  void out(float q)
  {
      cout << q << " is the value\n"
  }
```

```
PROGRAM outsub
    REAL :: x=3.14159 , y=2.71828 , z
    CALL out(x)
    CALL out(y)
    CALL out(x**2 + y − 0.3)
    STOP ' normal exit from outsub'
END PROGRAM outsub
SUBROUTINE out(u)
    REAL :: u
    PRINT * , ' The value is ' , u
    RETURN
END SUBROUTINE out
```

The calling program and the subprogram have to agree (at compile time) about the type and other details. In C++ the agreement is enforced using prototypes. **Fortran** has an ambiguous attitude: the traditional (i.e. as in earlier versions of **Fortran**) way is just to push the given argument list onto the stack, and hope that the subprogram will pop the right number, and agree about their types and kinds. This is clearly risky! A call to **out(100)** will place the integer representation of 100 onto the stack, which the function will then interpret as a floating point. A call to **mean(data)** omitting the second argument will cause the subroutine to pop two items off the stack even though only one has been put there for it. This is living dangerously. It's safer to provide the relevant details with an **INTERFACE** block.

A C++ main program's argument list is usually just empty, but it can be used to pass parameters from the command line that invoked it. If you use it this way then there are two arguments: an **int** which give the number of arguments, and a pointer to an array of **char** strings which contain them. The first string is just the name of the program as typed. So if your program executable is **myprog** and contains

```
main (int arg_count , char* args[ ])
```

and you invoke it with **myprog full** then the program will start with **arg_count** set to 2, **args[0]** pointing to **"myprog\0"** and **args[1] "full\0"**. This facility can provide an excellent means for the users to control what the program does.

5.2.1   Prototyping and interface blocks

With a **C++** function call the compiler has to know what arguments the function expects, and what will be returned. Given the function **mean** of §5.1, the statements

    int   j1 = mean(d , 100);
    float x2 = mean(d , 100);
    float q3 = mean(d , 100.0);

are all legal, but the first requires a float-to-integer conversion of the result of **mean**, and the second does not. In the third, the **double** 100.0 must be converted to the **int** 100 before it gets passed to the function. You make the information available with a *prototype*, as is done in §5.1 with the line **float mean(float data[ ] , int n);** right at the start of the program. The argument names are optional – they convey nothing to the compiler, and **float mean(float [ ] , int);** would have exactly the same effect, but they do make it more readable. It's handy to put such prototypes into header files* for convenience and uniformity.

Fortran has to know the type of any function return values, and so these are declared as **REAL** or **INTEGER** etc, as required. This is the purpose of the statement **REAL , EXTERNAL :: mean** in the program in §5.1. The 'external' attribute may be omitted in this instance as the compiler can deduce that **mean** is a function, but the use of **EXTERNAL** improves program clarity, which we strongly recommend.

The details of the argument list can be provided in an **INTERFACE** block at the start of the program, among the specification statements. This contains a copy of the subprogram statement and the declarations of its arguments

                INTERFACE
                    SUBROUTINE out(z)
                        REAL :: z
                    END SUBROUTINE out
                END INTERFACE

The interface block is an important component in **Fortran** programming; it provides all the details that the compiler needs to access a subprogram (which may be very complicated, or pre-compiled, or even written in another language). As well as checking the types, it also checks the rank of any arrays, and is crucial for keyword and optional arguments. We recommend its use. Perhaps it is not necessary in a small program, where everything is in the same file, but it's vital for larger programs spreading over several source files. It may happen that some of these argument uses will work even without the interface block, but relying on this is not at all advisable.

The two languages adopt different strategies if there is a clash between the argument given and the specification in the prototype/interface. **C++** will cast the given value to what it should be, as best it can. **Fortran** will give a compiler error. So if the argument to **out** has been specified as floating point, then **out(100)** will work in **C++** but give a compile-time error in **Fortran**.

---

* You can put the whole function definition in the header, but this is clumsy for functions of any length.

5.2.2   Fortran modules

For convenience these interface definitions can be kept in separate files, and read at compile time using the **INCLUDE** followed by the name of the file. This works in just the same way as the **C++** header files, though without the different types of bracket and default paths. But it's more effective to use modules and the **USE** statement.

Modules are a nice feature of **Fortran** that help keep control of subprograms and their interfaces. A module – and it makes sense to have one source file for one module – comprises a collection of subprograms, and their interface definitions are *implicitly included*. It can also contain type definitions and data that will be shared by the routines, and other interface information. A module can be tested by one program before use by another, and indeed may be made use of by many different applications.

A module is (presumably) kept in a source file of its own, and it is compiled separately. First come type definitions, variable declarations, and interface blocks; then, after the statement **CONTAINS**, subprograms can be given. The relevant information is made available to the compiler (how this is done is not specified: in a typical system the compiler writes a file with the extension **.mod** as well as the usual **.obj** or **.o** binary file.) Another program can invoke a particular module by the **USE** command: if you have a module called **graphics** then your program says **USE graphics** and the compiler reads the information in **graphics.mod**, or something to that effect.

```
PROGRAM analyse45
  USE graphics
  REAL :: x , y ! ...
  ! ...
```

The interface information for the subprograms in the module is made available automatically: you are not able to give it in an interface block as well.

Data can be shared between **Fortran** subprograms very neatly by using modules. The data you want to share is defined in the module, in the first part before the **CONTAINS** statement. It's then accessible to all the subprograms in the module: to share it with the main program you just insert the statement **USE** with the name of the module; this makes all the data defined at the head of the modules **graphics** available to the main program, as well as all the subprograms they contain. (This can be nested: modules can **USE** other modules.) Modules can also hold definitions of user-defined types that are then available to any program (or other module) that **USE**s them.

The **USE** statement is a bit like **include**, in that all the definitions in the module are available to the compiler as it proceeds through your program. It has some advantages, however. First, the code for the interfaces and bodies of the subprograms are in the same file. In **C++** one has to worry that the files – say, **graphics.h** and **graphics.cc** – are kept in step and compatible during development. Secondly, the modules are compiled first, so when the compiler meets a **USE** statement it will access the relevant information in an efficient compiled form, rather than going through the slow business of reading and parsing a text file.

Here is a module which can be used to produce PostScript graphics in **Fortran**. The module contains only two routines, to draw lines and circles, but it could readily be expanded.

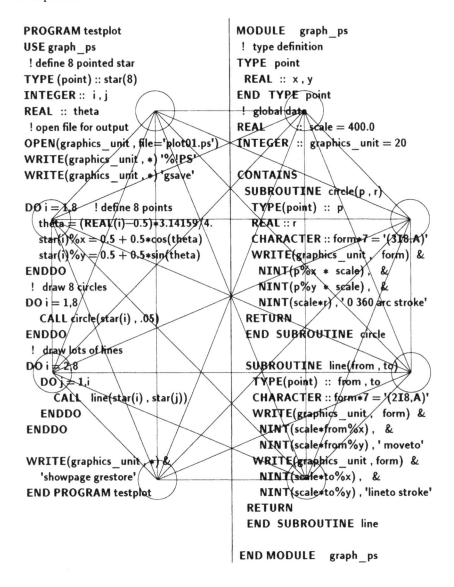

```
PROGRAM testplot                        MODULE    graph_ps
USE graph_ps                             ! type definition
 ! define 8 pointed star                 TYPE  point
TYPE (point) :: star(8)                    REAL  :: x , y
INTEGER :: i , j                         END TYPE  point
REAL  :: theta                            ! global data
 ! open file for output                  REAL     :: scale = 400.0
OPEN(graphics_unit , file='plot01.ps')   INTEGER  ::  graphics_unit = 20
WRITE(graphics_unit , *) '%!PS'
WRITE(graphics_unit , *) 'gsave'         CONTAINS
                                           SUBROUTINE circle(p , r)
DO i = 1,8    ! define 8 points            TYPE(point) :: p
   theta = (REAL(i)—0.5)*3.14159/4.        REAL :: r
   star(i)%x = 0.5 + 0.5*cos(theta)        CHARACTER :: form*7 = '(3I8,A)'
   star(i)%y = 0.5 + 0.5*sin(theta)        WRITE(graphics_unit , form) &
ENDDO                                          NINT(p%x * scale) , &
 ! draw 8 circles                              NINT(p%y * scale) , &
DO i = 1,8                                      NINT(scale*r) , ' 0 360 arc stroke'
   CALL circle(star(i) , .05)            RETURN
ENDDO                                     END SUBROUTINE circle
 ! draw lots of lines
DO i = 2,8                                  SUBROUTINE line(from , to)
   DO j = 1,i                               TYPE(point) :: from , to
      CALL  line(star(i) , star(j))         CHARACTER :: form*7 = '(2I8,A)'
   ENDDO                                     WRITE(graphics_unit , form) &
ENDDO                                          NINT(scale*from%x) , &
                                               NINT(scale*from%y) , ' moveto'
WRITE(graphics_unit , *) &                 WRITE(graphics_unit , form) &
   'showpage grestore'                         NINT(scale*to%x) , &
END PROGRAM testplot                           NINT(scale*to%y) , 'lineto stroke'
                                         RETURN
                                         END SUBROUTINE line

                                         END MODULE    graph_ps
```

Notice that the global variable **scale** is used by both routines in the module. The global variable **graphics_unit** is also used by the main program. In practice it would be neater to provide module routines to initialise and finish off the plotting.

5.2.3   Arguments: reference and value

There is an ambiguity in the passing of arguments which is present even at the most simple level. In a fragment like

x = 2.0;	x = 2.0
f(x);	CALL f(x)

does the subprogram access the variable **x** or the numerical value 2.0? The first is called call-by-reference, the second (which uses a copy of the variable's value) is call-by-value. If the subprogram is not intended to write to the variable, then this distinction doesn't matter, but if it is – and very often you do want to write a routine that manipulates its arguments – then it is vital.

The two languages take the two different approaches. In **Fortran** arguments are called *by reference*. The argument list for a procedure (a function or a subroutine) is a list of dummy values. If the actual argument in the call is a variable (including arrays, parts of arrays, or structures) then a pass by reference is made and the actual data can readily be modified in the routine. If the argument is an expression then that is evaluated and placed in temporary storage, and a reference given to that temporary storage.*

It is dangerously bad practice to change the arguments in a *function* whose apparent intent is to return a single result. Use a subroutine instead. Unexpected side effects of this kind should be avoided since they create complications in the mind. This leads to a classic Fortran bug as in, for example,

DO i = 1 , 10	SUBROUTINE fred(index)
CALL fred(i)	INTEGER :: index
PRINT * , i	index = index + 1
ENDDO	RETURN
	END SUBROUTINE fred

where the calling program (presumably) doesn't expect the index to be overwritten!

Such problems can and should be avoided by a helpful attribute called **INTENT**. When the dummy variable is declared in the subprogram (as it has to be to establish the type etc.) the **INTENT** attribute can be given as **IN, OUT,** or **INOUT**.

SUBROUTINE fred(index)
INTEGER, INTENT(IN) :: index
index = index + 1   ! Compiler will flag this as an error

The **INTENT(OUT)** attribute tells the compiler that the argument does not contain valid data on entry to the subprogram, so it will usefully fault any attempt to read data from it until it's been given within the subprogram. Strict use of **INTENT** for all arguments will save you hours of frustration at some future date.

---

* If the argument is a literal constant, 137.0337 say, it may be treated as an expression. Its value is copied to the procedure, and the original cannot be changed. But don't count on it!

In C++ the standard technique is to call arguments *by value*. Each argument is evaluated and the actual result is copied onto the stack, whether the argument is a constant **fred(3.4)** or an expression **fred(p*q)** or a variable **fred(x)**. The function, **fred**, is completely isolated from whatever invoked it. The copies of the values of the arguments exist only while the function **fred** is executing. Their scope is that function, and their duration (or lifetime) is that of the function. The advantages are that data is not changed (possibly corrupted) by the function, since only copies are used.

But you may want to do just that! A function may only return a single object, not even an array,* which is rather restrictive. To change an argument variable we can pass a pointer to that variable, instead of the variable itself. The pointer value is copied, but that copy can still be used to access the original for reading or for writing. The function writes/reads the result directly to/from the addresses without changing any of its own arguments.

There are two ways of doing this. One is more logical, the other more convenient. Locally the function – and the prototype – specify that an argument is a pointer by putting an asterisk in the argument declaration. The calling program must give it a pointer using the address-of operator **&** (or by giving an actual pointer variable).

Problem 5.2   *Consider the following two programs to swap two integer variables, and decide why* **swap1(&m , &k)** *works and* **swap2(m , k)** *doesn't.*

```
void swap1(int *p , int *q)              void swap2(int p , int q)
{     //  correct version             {      //  incorrect version
   int temp;                             int temp;
   temp = *p;   *p = *q;   *q = temp;     temp = p;   p = q;   q = temp;
}                                      }
```

The convenient way uses a something called a *reference*. For a given type (**float**, for example) you can declare a variable to be a reference to that type by using the ampersand character **&**.

```
float x;
float& rx = x;
```

This appears to be a point at which the language designers gave up any attempt to produce a logical/consistent syntax, in the interests of giving the users what they wanted. Don't try and understand that ampersand as an address-of operator. Just read the second statement as '**rx** is a reference to **x**'. A reference can be thought of as a pointer which can never be changed once initialised, and for which contents-of operator is implicitly applied whenever it is used. This has the effect that after the above declarations **rx** is loaded with the address of **x**, and when **rx** is used it gives the contents of the address of **x**, making **rx** an exact synonym for **x**: they are just two words for the same entity and you can use either interchangeably. That's not very useful. The point is that a reference can also be given in an argument list, for example,

---

* The object could be a structure containing an array, but this is running before we can walk!

```
float x = 1.234;                    void fun(float& rx)
fun(x);                             {
cout << "x value" << x << endl;         rx = 3.456;
                                   }
```

The **rx** 'pointer' holds the address **&x**, and the automatic 'contents-of' operation means that the function does the equivalent of **∗&x** = **3.456** and overwrites the actual **x** variable.

This is more convenient because, although the ampersand is specified in the function argument list and in the prototype, no special action need be taken either within the body of the subroutine or with the actual call – unlike the other method where you do have to take the address when you call and the contents when you use the arguments in the subroutine. So the routine

```
void swap3(int& i , int& j)
{
    int temp;
    temp = i; i = j; j = temp;
}
```

not only works, but also can be invoked just as **swap3(m , k)**.

Problem 5.3   *If a has been declared as an array of type* **int**, *does* **swap3(a[m] , m)** *work? What about* **swap3(m , a[m])**?

There are two reasons for using call-by-reference rather than the standard call-by-value. One is for overwriting arguments, the second is that, for a large object, call-by-value can be slow (and can overflow your stack). If you want to use call-by-reference for speed with an argument that you *don't* want to overwrite then you can specify it as constant.

void fun(const float& rx)...

Here again convenience scores over logic: this means not that the 'pointer' **rx** is constant: that's true anyway. It means that the contents of that pointer are constant, i.e. attempts to assign values with **rx** = **3.456**; will be trapped by the compiler.

5.2.4   Arrays as arguments

C++ treats arrays differently for they are automatically passed by reference. If you pass an array to a function in C++, what actually gets passed is the address of the first element, which is basically the same as a pointer: there is a close link between arrays and pointers (§7.2.4). Given **float array[100]**; then a call **float x = fun(array)**; matches the prototype/declaration **float fun(float a[ ])** or **float fun(float ∗a)**; with equal effect. Within the function members can be accessed as **a[j]** or as **∗(a+j)** as preferred.

A subtle point to watch is that **sizeof(a)** in the function returns the size of the pointer (probably 4 bytes) while **sizeof(array)** in the calling program returns the length of

100 **ints** (probably 400 bytes). Within the function the compiler doesn't know the size of the array, only that it is an array.

Given that a 1-dimensional array is much the same as a pointer, **C++** handles them easily. It knows the address of the first element, and as it also knows the size of the relevant type it can find the address of the second, third, fourth, and any other. It may access outside the array bounds but that is, as always, the programmer's responsibility. But for multidimensional arrays **Fortran** is clearly superior. To access, say, the word at the start of the second row: **a[1][0]** you have to increment the address of **a[0][0]** by **sizeof(a[0][0])** times the number of elements in the row – and the function has to know that. So if you give an array in a function definition argument list using square bracket notation you have to give the actual values for each dimension after the first:

```
void fun(float a[ ][10])        // correct
void fun(float a[ ][ ])         // wrong
```

If you now call **fun(array)** then **array** must be 2-dimensional, with the second dimension equal to 10. Otherwise the compiler will object.

This is very restrictive if you want to write general-purpose matrix routines. Matrix classes are around to supply just this deficiency and you may find one you can use – see also §7.2.6. If not, then you have to pass the necessary dimensions explicitly and do the array-access arithmetic yourself, switching from a 2-d array to a 1-d array of pointers, which is almost the same. But not quite. (See Kernighan and Ritchie (1988), section 5.9.) The program sums the array contents from **dd[0][0]** up to **dd[k−1][m−1]**.

```
float* dd[5];          // 1−d array of pointers         float func(float** a , int k , int m)
//  Not a 2−d array as in  dd[5][3]                      {
for(int i=0; i<5; i++) dd[i]=new float(3);                  float sum = 0.0;
dd[4][2] = 17.3;  //  elements accessible                  for(int i=0; i<k; i++)
                  //  in the same way                        for(int j=0; j<m; j++)
        /* ... */                                              sum += *(*(a + i) + j);
float total = func(dd , 5 , 3);                             return sum;
        /* ... */                                        }
```

Fortran saves us all this grief. An array passed as an argument to a subprogram must specify the number of dimensions, but that is all.

```
SUBROUTINE msub(a , b , c)
    INTEGER  :: a( : )                   ! a has 1 index
    REAL     :: b( : , : )               ! b has 2 indices
    LOGICAL  :: c( : , : , : , : )       ! c has 4
```

When invoked by **CALL msub(p , q , r)** the compiler will check that **p**, **q**, and **r** have the right number of indices and then call the routine; the necessary information about sizes is passed with it (by some mechanism we don't need to know about) and elements can be referred to within the routine.

The moral is that, whatever the superiority of **C++** in some areas, it is not a language well suited for serious array calculations, whereas **Fortran** is.

5.2.5 Checking arguments

When you program a function you presumably have an idea of the argument range, but another user may have different ideas. It is good practice to trap values which may cause problems such as overflows and fatal errors.

The lens formula in optics relates the object distance u from the lens to the image distance v by the equation

$$\frac{1}{f} = \frac{1}{u} + \frac{1}{v} \quad \text{or} \quad f = \frac{uv}{u+v}$$

where f is the focal length. Both values being infinite together is unphysical, as is either value being zero, but either may be negative. Suppose we write the function FOCAL, with two arguments, to find f given u and v, as in F_length = FOCAL(u , v):

```
REAL FUNCTION FOCAL(object , image)       ! returns the lens formula result
    REAL , INTENT(IN) :: object , image
    FOCAL = object * image / (object + image)
    RETURN
END FUNCTION FOCAL
```

All this is slapdash programming. Good programming practice strongly suggests that tests for the validity and range of arguments be made inside the subprograms we write, to protect them from the unpleasant consequences of an erroneous call. This makes them more independent and useful for their tasks. It pays to be paranoid. So in FOCAL the minimum test required is that the denominator be not zero.

This can be done by testing for zero and aborting the function evaluation with FOCAL = 0.0E0; PRINT * , ' FATAL ERROR... ZERO INPUT VALUE FOR u + v in FOCAL'. But what does 'testing for zero' actually mean? Real/float variables have a relative accuracy of about 1 in 10^7, i.e. EPSILON(u), so the sum is accurate to about $10^{-7}(abs(u) + abs(v))$. This is the basis of the 'test for zero' zerotest1:

```
LOGICAL :: zerotest0 , zerotest1 , inftest , totaltest     ! if .TRUE. abort
REAL :: u , v , zero=0.E0 , biggest=1.E30                   ! chosen cutoff value
zerotest0 = ( (u + v) == zero )                            ! better than nothing
zerotest1 = ( (u + v) <= EPSILON(u)*(ABS(u) + ABS(v)) )    ! big improvement
inftest   = ( ABS(u) >= biggest .OR. ABS(v) >= biggest )   ! data too large
totaltest = ( zerotest1 .OR. inftest )                     ! combines last two
```

The first and crudest test checks for precisely 0.0. The next traps all values whose reciprocals are close to overflow, and the third traps nasties at the other end of the scale; the two can also be ORed together.

This general principle can be applied to checking array bounds against data sizes, trapping negative or zero values for the real log and sqrt functions (before they are called), and so on. All of this subtlety may be hidden from the user. Anticipating problems, and dealing with them carefully, is a hallmark of good programming.

⌐ **5.2.6   Optional and keyword arguments**

The task performed by a subprogram can be a special case of a more general form. A simple average is a specific form of the weighted average, with all the weights equal to unity. The shift of a picture along the x axis is a specific case of a shift by an arbitrary vector (x, y). It is convenient to write one routine to handle all cases, but inconvenient to have to keep specifying a weight of 1.0, or a zero y-offset.

This is provided for in both languages. In **Fortran** the number of arguments in the call does not have to match the number of arguments in the subprogram definition. Within the subprogram the arguments which may be omitted are given the **OPTIONAL** attribute, and there is a function called **PRESENT** that specifies (at run time) whether the variable has been supplied or not.

```
SUBROUTINE total(value , weight)
    REAL , INTENT(IN) :: value
    REAL , INTENT(IN) , OPTIONAL :: weight
    REAL :: w        ! sumw and avg are REAL global variables
    IF (PRESENT(weight)) THEN
        w = weight
    ELSE
        w = 1.0
    ENDIF
    sumw = sumw + w
    avg  = avg + w*(value–avg) / sumw
    RETURN
END SUBROUTINE total
```

This can be called with two arguments or one, as desired. The arguments given are assumed to be in the order on the list; any omitted are at the end.

C++ always has all the argument values available in the routine, but provides a system of defaults so that you don't have to supply them all in the call. In the prototype you could write **void total(float value , float weight=1.0);**. Then if you use **total** with only one argument, the second will be set to 1.0. As with **Fortran**, if you leave out n arguments they must be the last n on the list. (Another way to provide this flexibility is by overloading – see §5.2.7.)

Fortran – but not C++ – provides the more flexible and friendly system of *keyword arguments* where the compiler uses the keyword to determine which argument is which, rather than the position in the list. If you want to add a value **x** with weight **w** to the total, but you can't remember offhand which of the two arguments to **total** is the value and which is the weight, you can call **total(value=x , weight=w)** or, equivalently, **total(weight=w , value=x)**. This is particularly useful where you have a routine with many parameters, and you only want to give some of them. Again, the **PRESENT** function can be used within the routine to discover whether an argument has actually been supplied. Interfaces (or modules) are essential when using these techniques: the **Fortran** compiler has to be told what the subprogram is expecting.

★ 5.2.7 Overloading

The same name can be used for a whole set of functions (or subroutines) which differ in their argument lists (number and type of arguments). An example is the standard **abs** function. Or, to be accurate, functions. There is a function in **<stdlib.h>** declared as **int abs(int)** which takes the absolute value of an integer and returns an integer. There is another in **<complex.h>** declared as **double abs(complex)** which takes a complex number and returns a double.* If the compiler encounters a call to **abs** it examines the argument given, and calls one or the other function depending on whether this argument is an **int** or a **complex**.

This works even with functions whose source is in some other file, and which are compiled separately and then linked to your program. This uses the cunning technique of *name mangling*: your function name (**fred**, say) in fact has added to it a string of unreadable bytes that describes the number of arguments and their type. This is why such external functions have to be declared with prototypes – usually by means of a .h file. The compiler then knows what functions are available, and whether the argument in a reference in your program to **fred(1)** should be left as an integer, as a function **fred (int a)** is available, or converted to a real for **fred(float a)**.

In **Fortran** the technique used for overloading is more explicit (it calls them *generic* procedures.) The subprograms are coded with different names. An **INTERFACE** block is then used to declare a generic name, within which details of the different subprograms and their arguments are given:

```
INTERFACE fred
    REAL FUNCTION    fred1(arg)
        REAL :: arg
    END
    REAL FUNCTION    fred2(arg)
        INTEGER :: arg
    END
END INTERFACE
```

and when the compiler meets a call to **fred(anything)** it will look at the options given in the **INTERFACE** block (or blocks) and implement it as a call to **fred1** or **fred2** as appropriate, depending whether the argument is real or integer in this case. If the precise type and kind are not available it will do the best it can.

The idea of overloading can be extended from functions to operators, particularly for handling user-defined types. The **C++** operators are defined through functions: **operator+(...)** defines the action of the + addition operator and **operator<<** the action of the left chevron. You can define your own versions of these for use with particular types. In **Fortran** the linkage is not quite so direct, but an interface can be provided to do this. §10.2 explains how this can be done.

---

* To find the absolute value of a float or double you don't use **abs** but a function **fabs**.

★ 5.2.8   Functions as arguments

Functions can be given as arguments to other functions. For example, you might want to write a routine to integrate any mathematical function numerically, using Simpson's rule, to be used as e.g. **float s = simpson(sin , 0. , M_PI , 100);**. In C++ the passed function, here **sin**, appears in the argument list, including its own type and its own argument list. (See also §7.2.8 for further information.)

```
double simpson(double f(double x) , double from , double to , int n)
{
    if(n%2 != 0) exit(1);              // n must be even
    float step = (to–from) / float(n);
    float sum = f(from) + f(to);
    for(int i=1; i<n; i+=2) sum  += 4.*f(from+i*step)+2.*f(from+(i+1)*step);
    return sum*step/3.0;
}
```

In **Fortran** when an argument is a function the calling procedure *must* specify the **EXTERNAL** attribute, §5.2 (unless it already knows from module information) as the compiler cannot deduce this from the context. Here is a program to find the zero of a function using a binary search, which in n steps locates the zero within $(high - low)/2^n$:

```
REAL FUNCTION seek(f , low , high , n)      PROGRAM testseek
  REAL , INTENT(IN)  :: low , high          REAL , EXTERNAL  :: fun
  REAL  :: lo , hi , mid , flo , fhi , fmid  !   this EXTERNAL is compulsory
  REAL , EXTERNAL  :: f                      !   for an argument which is a function
  INTEGER  :: i , n                          REAL , EXTERNAL  :: seek
  lo = low;  hi = high                       !   this EXTERNAL is optional
  flo = f(lo);  fhi = f(hi)                  !   but strongly recommended
  DO i = 1,n                                 PRINT * ,' zero at ', seek(fun, .01 , 3. , 20)
    mid = 0.5*(hi+lo);  fmid = f(mid)        STOP ' normal exit from testseek'
    IF (fmid*fhi < 0.0) THEN                 END PROGRAM testseek
      lo = mid;    flo = fmid
    ELSE                                     REAL FUNCTION fun(x)
      hi = mid;    fhi = fmid                REAL , INTENT(IN)  :: x
    ENDIF                                      fun = SIN(x) − 0.5*x
  ENDDO                                       !   SIN is an INTRINSIC function
  seek = mid                                 !   known to the compiler
  RETURN                                     RETURN
END FUNCTION seek                            END FUNCTION fun
```

Problem 5.4   *Make the above into a respectable function by (1) checking that the first two evaluations of* **f(hi)** *and* **f(lo)** *straddle zero and (2) taking appropriate action if* **f(x)** *is zero at any stage. You might also want a more sophisticated choice at each iteration than averaging the two values.*

★ 5.3   OTHER FACILITIES

★ 5.3.1   Template functions

The code (in §5.2.3) to swap two integers won't work for floats; for that we would need to rewrite it. And again for doubles, and again for any other type. A better way is a *template function* applying to any type. More details are given in §10.9.

```
template <class T>  void swap(T& a , T& b)
    {
        T  c;    c = b;    b = a;    a = c;
    };
```

This is a standard function definition preceded by the keyword **template** and an argument list in angular brackets: this specifies that in what follows **T** can be any sort of class ('class' here means a standard or user-defined type, structure, etc.). The template represents a set of functions: **swap(int& a , int& b)**, **swap(float& a , float& b)**, **swap(complex& a , complex& b)** and so on, and any call to **swap(p , q)** is matched against all possibilities.

★ 5.3.2   Inline functions

To declare a **C++** function *inline* is to ask the compiler to insert the relevant code wherever it is used. This avoids the overhead of a call and return, at the price of a larger object program. Use it only if you're really worried about timing, and if you think you're an expert – and anyway your request (like **register**) may be ignored by the compiler if it doesn't agree with your judgement. It should clearly be used if and only if the function is tiny, as is the case for the definition of the **abs** function

```
inline  int  abs(int d)
    {
        return (d > 0) ? d : −d;
    }
```

★ 5.3.3   Fortran internal subprograms

Fortran lets you have *internal subprograms*: they are enclosed between the **PROGRAM** and **END PROGRAM** statements using the **CONTAINS** keyword to show that the main program definition has finished.* Internal subprograms can't be invoked from anywhere outside the *host* program that encloses them. They have access to all the variables within their host, so can be used without having to make arrangements to pass arguments.† Any declaration within a subprogram is restricted to that subprogram. Examples of the usage are in §5.2.2 (in a module) and in §6.4.

---

* Separately compiled subprograms, and subprograms in a module can also have internal subprograms, but only one level of nesting is allowed: internal subprograms can't have their own internal subprograms.
† Convenient but risky as a carelessly written internal subroutine can overwrite in unthought-of places.

< 5.3.4   Sharing data

Arguments and return values are one means of communication. Global data (§2.1.3) is another which can be convenient when many routines need to access the same data. (For example, as in the module in §5.2.2 when several graphics routines all use the same variable for an I/O unit number.) But it is always potentially dangerous. One program could decide to change the unit number: the author of another might not realise this was happening. The danger can be mitigated by using non-obvious names to prevent accidents, by facilities for data-hiding, and by rethinking the program structure. For example, all the routines could write their output data to a character string, and then pass that string to a single routine which writes it to the output unit.

Global data can be used for the distribution of necessary information to subprograms. This is safer as the data is read, not written, so unexpected side-effects don't happen. It may be useful to declare a set of numerical and scientific constants relevant to the program. For example in C++ you will probably use some header files along the lines of

```
// This is   physics.h
const double PLANCK = 6.626E−34;
const double M_ELECTRON = 0.511;
const double M_PROTON = 938.3;
      /* ... */

// This is   myprog.h
const int MAX = 100;
const double TOLERANCE = 1.0e−20;
      /* ... */
```

```
// This is a 'typical' program
#include "physics.h"
#include "myprog.h"
main()
{
    float x_values[MAX] , y_values[MAX];
    float frequency = 3.0e14;
    float energy = PLANCK*frequency;
    cout << energy << " = energy\n";
      /* ... */
```

In **Fortran** you could compose a module called **physics** which you then pull in as required by a **USE** statement, and another one for the program parameters, using the **PARAMETER** attribute to ensure that they aren't overwritten.

The difference between these two uses is that you may well want to vary the value of **MAX** or **TOLERANCE** as your program develops; by providing it as shared data you ensure that you do so consistently. You're not going to change the value of Planck's constant; you're fed up with looking it up in reference books whenever you need it, so you do it once and get it right – and you make sure that it's specified in an obvious way in an obvious place: if the values provided aren't convenient then you'll go back to looking them up each time. This applies even more strongly if one author is providing values for a group software project.

Although the usual practices for the two languages, given here, look similar, there is a contrast between a value given in a **C++** **#include** file, where functions in separate files get different copies of the same thing, and one in a **Fortran** module, for which there is one copy to which separate units have access. The equivalent of the latter is the **C++** **extern** directive, as discussed in §2.1.3. **Fortran** also has an **INCLUDE** facility but modules are (we believe) a better way of achieving the same end.

★5.4 RECURSION

Sometimes you want a function to call itself. A classic example is the factorial function (here using long integers, §3.1.2). In C++ this presents no problems, except for the range. See more detail in the example program in §2.5.

```
long int function factorial(long int r)
{                    // note that 13! = 6,227,021,000 is already > long int
    return (r <= 1L ? 1L : r * factorial(r−1));
}
```

In **Fortran** recursion is also possible but you do have to go a little out of your way to do it. The function or subprogram must be specified as **RECURSIVE** and, if it is a function, the resulting return value can't be the function name (as the compiler can't then tell whether an instance of the function name is the return value or a recursive call) and so it has to be given some different name.

```
RECURSIVE INTEGER FUNCTION factorial(r)  RESULT(nfact)
    INTEGER :: r      ! the type of 'nfact' is declared along with 'factorial'
    IF (r <= 1) THEN  ! 0! is defined to be 1. Also trap negative values
        nfact = 1
    ELSE
        nfact = r * factorial(r−1)
    ENDIF
RETURN
END FUNCTION factorial
```

If you have a setup where function **A** does not call itself, but it does call function **B** which in turns calls **A**, then **A** still has to be specified as **RECURSIVE** but there's no problem with the return value name.

Problem 5.5   *This recursive factorial routine is extremely inefficient, and of rather limited use as integer overflow restricts it to a handful of arguments. Extend it to reals. Then compare its speed with that of a simple loop, using a 2-argument function which does not use recursion.*

A common use of factorials is in finding the number of combinations there can be of r objects drawn at random from a total of n. The formula is

$$^nC_r = n!/[r! \times (n - r)!]$$

and most of the terms cancel! Take the case of $N = 100$ and $r = 98$. Then $^{100}C_{98} = 100!/[98! \times 2!]$ Now observe that $100! = 100 \times 99 \times 98!$ so that $100!/98! = 100 \times 99$ and hence $^{100}C_{98} = 100 \times 99/(2 \times 1) = 4950$ by mental arithmetic.

Problem 5.6   *Code the above idea and evaluate ^nC_r for n=30, r=2, 15, 20 and for n=100, r=2, 15, 25, 98, 99, 100.*

★5.5   SYSTEM FUNCTIONS

Both languages contain standard libraries containing numerous resources. It is really useful to get to know the range and limitations of these libraries and how these affect your own work.

Fortran comes with a set of *intrinsic* functions and subroutines. You'll find a list in the documentation for your specific compiler, probably at the back, and a partial list in §9.5.2. The language standard specifies 108 intrinsic functions and 5 intrinsic subroutines, with purposes ranging from numerical computation (**SQRT, TAN, COSH , ...**) to bit manipulation (**IAND, IOR, NOT, ...**). Compiler vendors generally like to add a few more to lock you into using their non-standard flavour of the language.

These functions are always there and are no trouble. You don't need to memorise the 113 names, as if you do happen to give one of the names to a subprogram of your own, then that will have priority. A browse through the the relevant appendix is often fruitful, though, as it will tell you about features you might never have known about otherwise.

In **C++** a similar plethora of functions is available, but you have to know what you want in order include the appropriate header files (and if one of your own functions has the same name as one in an included header file, the compiler will object.) The back of your **C++** documentation will give you details of them all. For mathematics beyond simple arithmetic you need to **#include <math.h>** which contains sine, cosine, exponential and other routines (of which **pow** is one). **string.h** contains the string functions, and **time.h** the date and time. When these are included, the full name is given in angle brackets: this tells the compiler (actually, the pre-processor) that the file will be found in an *include directory* the name(s) of which is stored in some operating-system variable. (It's well worth finding this directory and having a snoop around to see what's available. It may also be helpful to look at the prototype when you're using one of these functions to check exactly what arguments it expects.)

★5.6   LIBRARIES

One of the keys to working productively and avoiding stress is the building up of a library of useful routines that can be applied to different problems, rather than having to start from scratch each time, just as one builds of a library of useful textbooks and references that one knows one's way around. Then when you're confronted with a new challenge, you've got a set of resources to call on, that you are familiar with and know your way around.

At the same time, as you write new code, it's worth bearing in mind that by writing a routine in a general way, rather than being specifically tailored to this peculiar problem, you can get an extra useful benefit from the project. It will probably help you write better code too.

So you should have your own modules, or **.h** and **.cc** files. Keep them in shape, and use them where you can. Although you won't intend to publish them, you will find yourself passing on routines to colleagues, so it does pay to keep them in the sort of tidy state which you wouldn't be ashamed for others to see. It's also vital to include comments explaining what things do. Although you may be totally aware of whether **scaleby(2.)** doubles or halves your picture size, and its clever effects on the global variables, you'll have forgotten all about it when you come back to change it in a year's time.

Projects written by teams of programmers are often based round a library of subprograms. The tasks of each subprogram are decided on and the interfaces between them are agreed. Then members go away on their own to write the individual components. Such libraries generally have long lifetimes, and become an established part of the analysis remaining long after most of the original programmers have departed.

Libraries of software are also available commercially. They are not cheap, and it can take you some effort to manipulate your problem into a form they can handle. But the good ones do contain the combined wisdom of many expert programmers, and it is often enormously quicker to use their subroutines rather than reinventing the wheel on your own. And your wheel is likely to be square the first few times!

6

Characters and Strings

Characters are treated first, and the way they can be handled. The ASCII code is presented, with notes on the way non-standard characters are used. Then the different approaches taken by the two languages to strings of several characters are outlined, and the supporting facilities that are provided are explained. Finally an example program is given to illustrate the methods which have been discussed.

An enormous number of computing applications deal with handling textual data, in English or other languages. These range from such mundane jobs as producing mailing-list labels to complex analyses such as identifying authorship from patterns in writing. There are many general and useful applications in between; for example prompting a user for a command, and understanding what they respond. This chapter considers this area, beginning with simple single characters and working up to the more complicated strings and substrings.

6.1 CHARACTERS

A byte of information in a computer, in memory or on disc, can represent not only a number (or part of one) but also a character (or part of one!). The value of a byte – there are 256 possibilities corresponding the integers 0 to 255 – represents a character. The representation in usually done using the *ASCII system*: details are discussed in §6.2.

In C++ there is a *character type*. Variables can be declared as type **char**, and a *character constant* is written as single character in single quotes. These variables and constants can be used in expressions just like 1-byte integers for many purposes. Thus **'A'+'a'** is a perfectly legal expression – it has the value 162 with the ASCII character set – and so is **'B'+1**. The C++ programmer is assumed to know about the character set they're using (ASCII or otherwise), and can do unobvious things like writing **xascii = x+'0'** to produce the ASCII value for a digit **x** in the range 0–9. This, however, does not make for clear or portable code.

In **Fortran** there is also a **CHARACTER** type, which appears similar – but this conceals a basic difference in philosophy. The **Fortran** compiler knows and cares about the difference between characters and integers, and expressions like **'B'+1** will be flagged as illegal. The **Fortran** programmer is prevented from doing this (for their own protection) but has the advantage that the program is independent of the character set used.

The C++ programmer has a choice: to use the numerical aspects of **char** and face a possible major rewrite on another system, or to treat **char** strictly as a character and use the header file **ctype.h** to decode the machine representation. We favour the latter approach, but we will also illustrate the former as (i) ASCII is heavily used in practice, (ii) it is often necessary to understand existing code, and (iii) it has an intrinsic interest as an example of how a character-set representation is implemented.

6.1.1 The source character set and execution character set

Both languages define a legal subset of characters which the compilers will accept. This *source character set* is part of the specification and is independent of the particular platform being used.

The source character sets contains the blank and

Fortran   a–z A–Z 0–9 , . ; : ? ! ' " ( ) < > / + − = # % & ∗ $ _

C++   a–z A–Z 0–9 , . ; : ? ! ' " ( ) [ ] {}< > | / \~+ − = # % & ^∗ _

Both languages treat _ (underscore) as a letter, and **Fortran** includes $ also.

Other characters (such as non-English letters) may be used in comments, and in character strings (and in data files) if they are defined in the *execution character set*, which may depend on the basic system and preferences set up when it was configured. Indeed the fine details probably will vary in this way as different languages need different alphabets (see §6.2.2).

6.1.2   Manipulating characters in C++

Characters can be cast to and from their integer representation (see §2.4.10), using the cast operators **char** and **int**, as in the example:

```
// Program to show conversion between int and char types in C++
#include <iostream.h>
main ()
{          // int <-> char conversion
   int in;
   char ch;
   cout  <<  "\n Please type a single character:  ";
   cin  >>  ch;
   cout  <<  "Symbol " << ch <<
       " is code " << int(ch) << endl;
   cout  <<  "\n Please type a number 32 – 128 ; ";
   cin  >>  in;
   cout  <<  "Code " << in <<
      " is symbol " << char(in) << endl;
   return 0;
}
```

C++ provides a useful set of functions, Table 6.1, made available by **#include <ctype.h>**, that can be used to test the nature of a character. One of these is **isprint()** whose value is **true** if the character is printable (and this includes a blank). These functions know whether the ASCII system is used or not. An example is given in the next program (which uses **printf()**[*] rather than **cout**):

```
#include <ctype.h>
#include <stdio.h>
main()
{
   char  ch = 0;
   for (int k=0; k<256; ch = char(k++))
     {
     if ( isprint(ch) )
        printf("%d = hex %x is printable as %c \n" , ch , ch , ch);
     else
        printf("%d = hex %x  is not printable \n", ch , ch);
     }
}
```

---

[*] In **printf()** print formats are directly included. See §8.1.1.

Table 6.1 C++ logical functions available in **ctype.h** for examining the character **c** and testing its nature. Note that **c** can be a number of type **char** or **int**, but not a string

Function Name	Returns 'True' (i.e. non-zero) if...
int isalnum(c)	c is a letter or a digit
int isalpha(c)	c is a letter
int iscntrl(c)	c is a control character
int isdigit(c)	c is a digit (0–9)
int isgraph(c)	c is a printable, non white-space, character
int islower(c)	c is a lower-case letter
int isprint(c)	c is a printable character (including blank)
int ispunct(c)	c is a punctuation character
int isspace(c)	c is a space, form feed, newline, cr, tab, vertical tab
int isupper(c)	c is an upper-case letter
int isxdigit(c)	c is a hex digit

The **ctype.h** file also contains the function **tolower(c)** which takes a character and returns it, converted from upper to lower case if appropriate. Likewise **toupper(c)** converts to upper case. Suitable (in)action is taken when c is not a letter: **toupper('5')** gives **5** (not **%**, as it might appear from many keyboards).

6.1.3   Manipulating characters in Fortran

In **Fortran** a set of functions provides the facility to convert between the numerical codes and the corresponding characters. These are given in the following table.

Table 6.2   **Fortran** functions for converting characters to integers, and vice versa

Function Name	Argument Type	Return Type	Value
ACHAR(n)	INTEGER	CHARACTER	character corresponding to **n** in the ASCII character set
CHAR(n)	INTEGER	CHARACTER	character corresponding to **n** in your system's character set
IACHAR(c)	CHARACTER	INTEGER	integer corresponding to c in the ASCII character set
ICHAR(c)	CHARACTER	INTEGER	integer corresponding to c in your system's character set

Usually **CHAR** and **ACHAR** will be identical, there will only be a difference between them if your system is one that doesn't use the ASCII character set.

Here is a sample program to illustrate the **Fortran** character representation, using the hex formats (Z...) in **Fortran** (see Table 8.1 and Appendix 2).

```
PROGRAM charint
  ! Conversion integers <—> printable characters in the F90 ASCII character set
  INTEGER     :: int
  CHARACTER  :: char*1 , fmt*40 , label*40
  label = '( 6(1X , A) )'
  PRINT  label , 'Code (dec)' , 'Code (hex)' , 'Symbol' , &
        &'Code (dec)' , 'Code (hex)' , 'Symbol'
  fmt = '( 2(1X , I8 , Z11 , 7X , A1 , 1X) )'
  DO int = 32 , 127
      char = ACHAR(int)
      PRINT fmt , int , int , char , int+32 , int+32 , ACHAR(int+32)
  ENDDO
  STOP
  END PROGRAM charint
```

The output contains the following:

Code (dec)	Code (hex)	Symbol	Code (dec)	Code (hex)	Symbol
61	3D	=	93	5D	[
62	3E	>	94	5E	^
63	3F	?	95	5F	_
...					

★ 6.1.4   Gaps and tabulation

C++ allows you to include some formatting instructions as non-printing characters in your text strings. It does this by means of *escape sequences*. The *backslash character* \ is treated specially, and taken together with the next character, or characters, in the program source to give a single character in the running program. We have already met '\n'. Likewise '\t' gives a tabulation skip. You may try **cout** << " abc\tdef"; for a demonstration.

The complete list of such characters is \a = **alert(bell)** , \b = **backspace** , \f = **formfeed** , \n = **newline** , \r = **carriage return** , \t = **horizontal tab** , and \v = **vertical tab**. The most used, by far, is \n. Complete generality is obtained by the sequences \ooo and \xhh, where ooo are three octal digits and hh two hex digits. To get the backslash character itself you use \\, and '\0', known as NULL, can be useful: it gives a **char** which has the value zero. (See §8.2.)

Fortran specifies layout with the format instructions, not with the string. There are several horizontal tab instructions: (i) **Tn** to position **n**, (ii) **TLn** left **n** positions, (iii) **TLn** right **n** positions, (iv) **nX** right **n** positions. So try as an example **PRINT '(A , T1 , A)' , 'abd' , 'def'**. A vertical tab, as such, is absent but the slash / terminates a record and starts a new line. **k/** moves **k** lines down. (See also §8.2.)

★ 6.1.5   Control characters; operating system differences

Many computing systems are DOS-based or UNIX-based. This difference in operating systems is more than just different names for the same operation – such as **dir** or **ls** to list the files in a directory. It also affects what you produce when you press 'Return' on the keyboard (sometimes the key is labelled 'Enter' or 'Newline'), or output the **\n** character. This gives the byte-pair **0xD , 0xA** = **\n** (also called CR and LF) on DOS systems, but on a UNIX system only **0xA** appears.

In the mists of antiquity people used manual typewriters, and went from one line to the next in two steps, first returning the carriage, which held the paper, to the first typing column by slamming the thing sideways, and then feeding the paper one line upwards by nudging a ratcheted lever. Some systems keep these meanings, using CR to return to the first column on the same line, and LF to move to the next line down in the same column, and such a system (e.g. DOS) needs both to start a new line. Other systems regard this as archaic, and use only one code for 'Newline'. Thus in UNIX text files lines are separated by a single LF character, whereas in DOS the two-character pair CR–LF is used.

If a file prepared under UNIX is printed by a DOS system, at the end of each line of text the cursor moves one line down but does not revert to the left-hand column, and the output is spread out in a falling cascade and soon disappears off the right-hand side of the screen. In many cases conversion is easy; read a UNIX file into an intelligent DOS editor and SAVE the output; all LFs become CR–LFs and vice versa. With a stupid editor (or an ignorant one – how does it know that this is a UNIX file?) there will be problems.

Likewise your printer may or may not generate an automatic line-feed when it is sent a CR character – this is usually controlled by some setting-up procedure. If you get this wrong your listing will be produced with unwanted double spacing or, even worse, no line spacing at all. This is generally handled by friendly local software: if a C++ program outputs the **\n** newline character (which is ASCII 10=**0xA**) on a DOS system it actually writes two bytes, 13=**0xD** and 10=**0xA**. On input the reverse occurs. You only need to worry about this if you're doing complicated things involving binary input or output – for example in a screen-dump program where a bit map of binary data is fed to a printer; some of the bit patterns will have values 10 or 13, and the 'helpful' printer software is liable to remove them, or add extra characters, and ruin the picture.

Problem 6.1   *Files can be transported between computers across networks using* **ftp** *(File Transport Protocol). A command to use this is available on many systems. This provides a way (in some setups it will issue a prompt) for the user say whether the file being transferred is text or binary. Why does it need to do this? What are the consequences of giving the wrong type?*

★ 6.2   THE ASCII CODE

ASCII is the American Standards Committee for Information Interchange, and the ASCII code is their particular mapping of characters to integers. These sections delve into the mysteries. In earlier times all programmers needed some of this detail, but now it is less necessary.

In the ASCII table the value 65=**0x41** represents the character 'A', 'B' is 66=**0x42**, and so on all the way to 'Z' at 90=**0x5A**. The lower-case letters 'a' to 'z' are represented by the numbers 97=**0x61** to 122=**0x7A**; each lower-case letter has a value 32=**0x20** more than the corresponding upper-case one. Most serious computing books contain a table of the ASCII character set, and this is no exception (Appendix 2). Printing a character from a program involves taking a byte of data (say 88=**0x58** , ASCII 'X'), and sending it as an 8-bit signal to some device, such as a screen or printer. This hardware then depicts, in its own way and its own style, the character 'X', either directly or, if the output device is bit mapped, using a font table.

As well as the 52 letter-codes, the ASCII system deals with other characters: the digits 0 to 9 are represented by the codes 48=**0x30** to 57=**0x39**, there are various punctuation marks and special symbols – comma, brackets, ampersand etc.

The integer value corresponding to a particular character is called the *code* for that character. It is important to distinguish between numerical *codes* and numerical *characters*. The character 7, for example, has ASCII code 55=**0x37** whereas the ASCII code 7 is a non-printable character ('alert/bell'). If you understand this difference then you understand the principle of character representation.

★ 6.2.1   ASCII control characters

Characters with codes below 32=**0x20** are called *control characters*. They do not produce visible output characters, but have other effects. To help talk about them, ASCII have given them all two- or three-letter names, and to input them from the keyboard the Control key is used. For example, ASCII(7=**0x7**) is known as BEL; it can be input* as Control-G, and on output usually makes an audible beep.

ASCII(8) is BS for backspace, can be input as Control-H, and should move the cursor back one space. Whether it actually does, and whether it erases the previous character, depends on the output device. Other control codes are more obscure, with different effects for different systems and devices. Further confusion is added in IBM PC systems, which use these codes for new characters such as smiley faces. You can see what they do on your system by experimenting, cautiously, with statements like **cout << char(8);** or **PRINT** ✳ **, CHAR(8)**; though beware that some control characters may do drastic things, such as disabling subsequent output.

ASCII(27=**0x1B**) (ESC – escape) is another interesting character. It is used for sending specific instructions detailed in characters that follow. For example, if your printer is controlled by the HP-PCL protocol, then if you send it the string of 5 bytes 27–40–115–49–83 then they themselves will not be printed, but the subsequent

---

* The Control key subtracts 20_{16} = **0x20** = 32_{10} from the usual value. G is **0x27** so that Control-G = **0x7**.

characters you send will be printed in *italic type* until you send the sequence 27–40–115–48–83 to return it to normal. Of course, if your printer doesn't use this protocol then this won't work (and goodness knows what will actually happen), but there will probably be some sequences that do the same job. Details of such fascinating ways of making printers produce different styles of output are usually in the appendices at the back of the manual. A *Printer Driver* is a program which prints data using the appropriate sequences for a particular printer.

Problem 6.2   *What is the connection between the* **C++** *escape character* \ *and the ASCII ESC character?*

★ 6.2.2   Character sets, fonts, and styles

Although most of the ASCII codes below 127=**0x7F** are well-established standards, there are some minor differences around, and for codes 128–255=**0x80–0xFF** there are many variations. In different character sets a given value may correspond to a totally different symbol. For example there is a character set designed for German use: most values are the same as standard ASCII, but about a dozen of the more obscure characters are used to represent the characters of the German alphabet which are not found in English: code 92=**0x5C**, the backslash symbol \ in normal ASCII, corresponds to Ö in the German character set. Likewise there is a Norwegian set where 92 gives Ø, and one for Spanish where it gives Ñ. **Fortran** allows in principle for different character sets within the same program (and even for characters of size other than one byte, for use with systems of writing using many symbols, such as Chinese). It is then possible to have statements like **CHARACTER (KIND=GERMAN) message∗20** provided your **Fortran** system knows about this particular alphabet.

ASCII(35=**0x23**) can be a problem. It is usually used for the character # and is known in the US as 'pound' (weight) but in some character sets it is used for the UK pound sterling £. For typesetters it is the 'octothorp' (count the points), for musicians it is the 'sharp', and in computing circles it is usually known as 'hash'.

A *font* is a way of representing a symbol. Although the value of, say, 65=**0x41** is defined as the character 'A', the ASCII standard does not prescribe anything about the shape of the character. It might be printed 'A' or 'A' or 'A' or in some other way. The font (or fount) specifies this.

Finally the *weight*, *style* and *size* specify different ways of printing the same font – in bold weight or italic style, for example, and in an enlarged or reduced size.

Changes of such details for simple printers and (insofar as this is possible) VDU terminals are normally done by ESC sequences. (PostScript printers are different.) For simple memory-mapped display devices there will be particular hardware components that determine the screen format, and device-specific commands will be used to control them. For more complicated bit-mapped devices the output is done through a driver program which reads the details of the fonts used from separate font files.

6.3 STRINGS OF SEVERAL CHARACTERS

Characters are often collected together to form words, sentences, names of people, mailing addresses, chemical formulae, or lines of text. Such a collection of a number of characters, in order, is generally referred to as a *string*.

6.3.1 String constants

String constants are set up in the program and do not change while it runs. We have met these before and used them frequently in output. Statements like

```
cout << "hello, world\n";          PRINT * , ' hello, world'
printf("hello, world\n");          WRITE(* , *) ' hello, world'
```

print the string **hello, world**. In C++ the appropriate header file must be loaded by **#include <iostream.h>** for **cout**, and **#include <stdio.h>** for **printf**.

Some small and niggling differences between them can be seen: C++ uses double quotes for strings in contrast to single quotes for characters, whereas in **Fortran** single or double quotes can be used to delimit a string. The **\n** is inserted in the C++ program because the **Fortran** print statement automatically puts a new line at the end of the output, whereas the C++ output statement doesn't. And finally, note the leading space in the **Fortran** string. This is to send a benign, do-nothing, character to the printer/screen, for some other characters have their own consequences. For example on some systems a '0' advances an extra line and '1' skips to the next page.

The C++ escape sequences can be used in strings, just as with single characters. These can also be used to include the demarcation characters '(apostrophe) and " (speech marks) in a string, using the sequences \' and \". A string like **"abc"def"** will give an error – neither compiler is smart enough to match the first and final double quote, and realise that the one left over must be intended as an actual character. In **Fortran**, as there are two alternative string delimiters, you can use one to include the other: **'abc"def'** and **"abc'def"** are both acceptable: if you really have to include the delimiter you're using, you just double it: to print **can't** you specify **'can' 't'** with two consecutive apostrophes.

You may want to use very long strings (longer than the width of the programming page). In **Fortran** this can be done using the continuation character, **&** at the end of the line. In C++ there is no problem writing multi-line statements since the only terminator is the ;, and it's also useful that adjacent strings get stuck together by the compiler:

```
cout << "hello hello hello, "        PRINT * , ' hello hello hello, &
"you big wide wonderful world\n";      & you big wide wonderful world'
printf( "hello hello hello, "        WRITE(* , *)' hello hello hello, &
"you big wide wonderful world\n");     & you big wide wonderful world'
```

The **&** at the start of a line in **Fortran** defines precisely where the string continues.

6.3.2 String variables

Now we consider string variables – collections of characters which we wish to treat symbolically. In C++ this is done by means of an *array* of type **char**: a collection of one-byte integers consecutive in memory. There is thus nothing different or special about a string as far as the C++ compiler is concerned – it's just an array of (rather small) numbers, and the size of the array determines the number of characters that can be stored in the string. In **Fortran**, on the other hand, although the string is also a set of consecutive bytes, it is declared as being of a specific character type, and the string length is given. **Fortran** does not handle single characters separately, but treats them as strings of characters of length 1; the declaration **CHARACTER** c used earlier is equivalent to **CHARACTER** c∗1.

So to specify that you want a string called **alphabet** to be used to contain the 26 letters A–Z, you would say

char alphabet[27]; | CHARACTER   :: alphabet∗26

(There are several ways of specifying the length in **Fortran**. We prefer this one and will stick with it.)

Why does C++ need 27 locations for 26 letters? Because – and this is a fundamental point which will be discussed in more detail later – a C++ string contains not only the characters you see, but also a termination character \0 that marks the end of the string, and space must be allowed for it.

Problem 6.3   *Explain why the expressions* **'x'** *and* **"x"** *have the same significance in* **Fortran**, *but are very different in* C++.

Individual characters in a C++ string can be accessed as members of the array. When doing so remember that *arrays in* C++ *are indexed from 0*. So the 27-long string **alphabet** can be used to contain the 26 letters of the alphabet starting with A as element 0 and Z as element 25. (Accessing individual characters in **Fortran** is done using substrings, and will be described in section §6.3.8.)

A string can be initialised in the declaration if desired, and constants, or constant expressions, can be used in specifying the size. (The brackets in **Fortran** are necessary when giving the length of a string with anything other than a simple number.)

const int MYSIZE = 10; | INTEGER, PARAMETER::SIZE = 10
char message[MYSIZE] = "Syzygy"; | CHARACTER::message∗(SIZE)='sizigi'
char longermsg[2∗MYSIZE+10]; | CHARACTER::longermsg∗(2∗SIZE+10)

When you pass a string in an argument list, the subprogram receiving it must know what it has, so you need to declare it there as being a string. However you don't need or want to give the size, because this secondary declaration will not cause the compiler to allocate space for it – that's been done already in the calling program. The length is thus omitted entirely in C++: **char message[ ]** and replaced by an asterisk in **Fortran**: **CHARACTER :: message∗(∗)**.

A useful application of single character handling is in the little routine which follows, which is a useful thing to keep in one's library(§5.6). Very often in controlling a program one wants to ask the user a question with a simple yes-or-no answer (such as **'Do you want instructions'** or **'Is there any more data'**). This can be handily done by this function; the call will print the question, and return the value TRUE if the user types **Y** or **y**, and FALSE if they type **N** or **n**. If they type anything else the question is repeated until an intelligible answer is obtained.

A program that uses this should contain the appropriate declaration: **LOGICAL query** for **Fortran** and **int query(char q[ ]);** for **C++**. In C++ this will probably be placed in your private library header file.

```
int query(char q[ ])
//  Library function that prints a query
//  obtains a yes—no answer
//  and returns true or false
//
//    See section 6.1.2 for 'toupper'
{
  char reply;
  while (cout << q)
  {           // repeat till satisfied
    cin >> reply;
    switch (toupper(reply))
    {
      case 'Y':
                   return 1;
      case 'N':
                   return 0;
    }
    cout << "Eh? Please reply Y or N\n";
  }
}
```

```
LOGICAL FUNCTION query(q)
! Library function that prints a query
! obtains a yes—no answer
! and returns .TRUE. or .FALSE.

CHARACTER :: q*(*) , reply*20
DO WHILE (.TRUE.)
    PRINT * , q
    READ * , reply
    SELECT CASE(reply)
    CASE('Y'); query = .TRUE.
    CASE('y'); query = .TRUE.
    CASE('N'); query = .FALSE.
    CASE('n'); query = .FALSE.
    CASE DEFAULT
        PRINT * , ' Sorry?' ,&
        &' Please reply Y or N'
    CYCLE
    END SELECT
RETURN
ENDDO
END FUNCTION query
```

Note that the **C++** program acts only on the first character and will leave two others unread should the response be **yes** rather than **y**. **Fortran** reads the string and if this is more than one character it will never match.

Problem 6.4   *Change the* **C++** *program to remedy this defect by reading the reply as a character string and then looking at the first character.*

6.3.3   Strings, arrays, and pointers in C++

The **C++** language allows you much closer to the heart of things, and this is particularly true in its treatment of strings and arrays.

A string is an array of type **char** terminated by a null – nothing more and nothing less. If you declare

```
char ichar;
char message[10];
```

then a particular character – for example **message[3]** – is a variable of type **char** and can be used in exactly the same way as **ichar**. The two are syntactically identical. But what about the variable **message**? It can be used, in subroutine calls and **cout** statements for example, and in some way it refers to the whole string – how does it do that?

From the compiler's point of view, **message** refers to an area of storage somewhere in memory, and the declaration instructs it to set aside 10 bytes there. Suppose these start at an address of **0x123456**. Then, when it encounters **message** later in the program, it refers back to that value, so **cout << message;** means 'send the char array starting at **0x123456** to the output stream'. Such an address is called a *pointer*. Pointers are a separate type of variable, and form an important topic of their own. (See §3.6 and Chapter 7.) They can be declared as variables in their own right, for instance

```
char *pmessage;
```

which declares a variable called **pmessage** as a *pointer to type char*. (The asterisk means it's a pointer.) This is 4 bytes long (probably – the exact length depends on the system), and the **char** in the definition tells the compiler that what it points to is to be treated as a variable of type **char**.

You can get at the object pointed at by using the asterisk *operator* which means 'the contents of'. Thus ***pmessage** in a program means 'the object that **pmessage** points to', which will in this case be treated as a character. So in a construction like

```
pmessage = message;
cout << *pmessage;
```

the character in **message[0]** will be printed. By contrast, **cout << pmessage;** prints the whole string.

Further characters in the string can then be accessed using the pointer, so ***(pmessage+3)** refers to the item 3 up from **pmessage**. The effect of square brackets can be understood in the same way. **message[3]** and ***(pmessage+3)** and **pmessage[3]** and ***(message+3)** are equivalent.

The opposite of the 'contents of' operator ***** is the *address-of* operator **&**. The result of this operator is a pointer. So **&message[0]** is the same as **message**.

The similarities and differences between an array name (on its own) and a pointer are a profound matter: a suitable subject for meditation by Buddhist monks on lonely mountain tops. Both refer to an address in memory, and can be treated in syntactically similar ways in many cases. For example, an array of **char** can be passed to a subroutine using **char f[ ]** or **char *f**. The sixth member can be referred to as **f[5]** or ***(f+5)** – notice the brackets in the latter case, as the operator precedence makes ***f+5** into **(*f)+5**. However a pointer can be assigned an expression, whereas an array name cannot: **message = "hello"**; is legal if **message** has been declared as **char *message** but not if it has been declared as **char message[10]**. On the other hand, **message[3] = 'x';** and ***(message+3) = 'x';** are always legal if **message** has been declared as an array, but if it is a pointer it must also have been assigned to some suitable value, so that the character is put somewhere sensible in storage.

Another difference arises with the **sizeof** operator. **sizeof(message)** returns the size of **message**, as declared. (Or the number of bytes – same thing for type **char**.) **sizeof(pmessage)**, on the other hand, always returns the number of bytes in a pointer (usually 4). Incidentally, if an array **arr** is passed to a subprogram, within that subprogram **sizeof(arr)** will always return this value.

6.3.4   String expressions

Two strings may be joined in **Fortran** using the *concatenation operator* `//`. In C++ it is not quite so easy, but there is a function **strcat(to , from)**, made available by **#include <string.h>**, which takes the second string and copies it onto the end of the first. (There is also a function **strncat(to , from , n)** which copies a maximum of **n** characters, although `'\0'` causes trouble.)

char forename[10] ,
     surname[10] , fullname[20];
 strcpy(fullname , forename);
 strcat(fullname , surname);
	CHARACTER :: forename*10 , &
	     surname*10 , fullname*20
	 fullname = forename//surname

Fortran supplies some other routines which manipulate strings to give strings.

Table 6.3   **Fortran** string functions

Function Name	Returns
REPEAT(s , n)	**s** repeated **n** times
TRIM(s)	**s** with trailing blanks removed Note that if blanks are removed the resulting string is shorter than **s**
ADJUSTL(s)	**s** with any leading blanks removed and trailing blanks added
ADJUSTR(s)	**s** with any trailing blanks removed and inserted at the beginning

★ 6.3.5   Arrays of strings

One frequently needs to work with a whole collection of strings – lists of names, for example, or sets of possible responses to a query.

In **Fortran** this is very straightforward. There can be arrays of strings, just as there can be arrays of integers and arrays of reals. The set of names of the chemical elements in the periodic table can be declared as

> CHARACTER   ::   symbol(92)∗2 = (/'H ' , 'He' , 'Li' , 'Be' , 'B ' , ... /)

The strings can be accessed as whole entities – **symbol(3)** would be **'Li'** if we store them in sequence, and treated in the usual way **symbol(3)(2:2)** will be the single-character substring **'i'**. All you have to remember is that the array index comes before the substring range.

In **C++**, as a string is itself an array, a one-dimensional array of strings is just a two-dimensional array of characters. So the equivalent array is (remembering that in **C++** each dimension has a separate square bracket)

> char symbol[92][3] = {"H " , "He" , "Li" , "Be" , "B " , "C " , ... };

These strings can then be referred to individually: if the names of the elements have been stored in order then **cout << symbol[1]**; will produce the output **He**.

The order is again array index first, string index second: in **C++** multiple dimensioned arrays are stored in memory with the last index varying most rapidly, so that the element **array[1][1]** is followed by the element **array[1][2]**. (This is the opposite from **Fortran**, in which **array(1 , 1)** is followed by **array(2 , 1)**.)

6.3.6   String assignments

Having got a string you want to use it – for example by copying one string to another. In **Fortran** this is straightforward; you might have, for example

> CHARACTER warning∗30, message∗30
> warning = 'Negative Root'
> message = warning
> PRINT ∗ , message

This does what you would expect. There are not enough characters to fill the **warning** variable in the first assignment statement, but this does not matter: the extra characters at the right are padded out with blanks. (If the target string is too short for the string assigned to it, characters are removed from the right, and no error message is given!)

In **C++** things are not quite so easy. You can't say

> char warning[30] , message[30];
> warning = "Negative Root";   // ILLEGAL
> message = warning;   // ILLEGAL
> cout << message;

To understand why not, remember that string names are equivalent to pointers.

When the compiler encounters a string constant (**"Negative Root"**) it stores the string, including the final null terminator, somewhere convenient in memory, and syntactically what appears in the program is the address of the start of the stored string – say, **0x123456**. Likewise the name of a string in the program is the address of the start of the space reserved for that string in memory – say, **0x456789**. So the apparently innocuous statement **warning** = **"Negative Root"** is tantamount to saying **0x456789** = **0x123456** which is a nonsense – you might as well try saying **1** = **2**. *This is particularly confusing as such 'assignments' are allowed in an initialisation where the syntax looks similar but is really different: an initialisation statement is not the same as an assignment statement!* So the sequence

```
char message[10]="hello";
```

is fine and legal, but not

```
char message[10];
message = "hello";
```

There are two ways round this. One is to declare the variables not as strings of characters but as *pointers to characters*. It is correct to say

```
char *warning , *message;
warning = "Negative Root\n";
message = warning;
cout << message;
```

If you want to work with strings without getting into pointers, then there are several standard functions provided which can be used. Their declarations are contained in the header file **string.h**. In this case we can use **strcpy** which copies one string to another, as follows:

```
char warning[30] , message[30];
strcpy(warning , "Negative Root\n");
strcpy(message , warning);
cout << message << endl;
```

which outputs **Negative Root**.

There is a function **strncpy(to , from , n)** which is similar to **strcpy**, but which copies only a certain number of characters (or until a null is encountered first). Note that this can often result in no null-termination to the new string. Also **memcpy(to , from , n)** and **memmove(to , from , n)** can be used to copy n characters, regardless of whether any nulls are encountered. (The difference between them is that **memmove** works correctly even if the strings overlap.) Thus

```
char timeofday[30]  =  "Good what?";
char message[30]  =  "Each morning!";
strncpy(message+5 , "eve" , 3);
strncpy(timeofday+5 , message+5 , 7);
cout << timeofday << endl;
```

produces "Good evening".

6.3.7 String input and output

The two languages handle input and output in radically different ways. For **C++** a device produces or absorbs a *stream* of bytes, one after the other, whereas for **Fortran** a device deals with *records*. The two approaches go back to the old days when input came either on paper tape, which was just a stream of bytes, or on punched cards, for which every card had 80 characters.

This is why, in the simple output statements we have already met, the newline character must be explicitly specified for **C++** (either by **\n** or **endl**), but not for **Fortran**

cout << string << endl; | PRINT * , string

as **C++** just sends the bytes in **string**, up to but not including the terminating null, to the output. In **Fortran** this writes a record containing the characters of **string**, and a record (on standard output) is terminated by a new line.

More complicated forms can also be used in a sensible way:

cout << "Error! " << x << " is < 0.0 \n";| PRINT * , ' Error! ' , x , ' is negative'

If you want to specify the output format, when printing tables, for example, **Fortran** uses the format statement:

PRINT '(A6)' , string

will write 6 characters. If **string** is longer than this field width of 6 then the first 6 characters are used, if it is shorter then extra blanks are added. The equivalent functionality in **C++** is obtained by using the *field width manipulator* **setw**, which requires you to include the header **iomanip.h**. When sent to the output stream this sets the width of the next item, so

cout << setw(6) << string << " has occured \n";

will pad the length out to 6 characters even if **string** is shorter. However if it is longer it is (unlike **Fortran**) not truncated. This only affects the next item sent, so " **has occurred \n"** is not affected.

The function **cout.write(string , n)** writes **n** characters from the string, irrespective of the presence or absence of any null characters. There is also a function **cout.put(c)** that outputs a single character.

Input of characters is similar, though a little more involved, as you have to specify when you want to stop.

READ * , string

will read **string** from the keyboard (assuming it is unit 5). If the string contains a blank, it has to be enclosed in quotes (single or double). If the input string is too long for **string** it will be truncated at the right, if too short it will be padded.

READ '(A6)' , string

is similar, except that it will read 6 characters from the record, be they blank or non-blank, and put them in **string**, truncating or padding as necessary.

In **C++** it is important to note that the standard input method **cin >> variable;** has the useful but occasionally confusing property that it skips over leading white-space characters. (This can be altered if required using I/O manipulator flags.) If you want to read a character **c** which may be blank (or tab, etc.) then you have to use **cin.get(c)** instead.

For string input, the simple input **cin >> message;** works fine, with leading blanks suppressed. You can also use **cin.get(message , 10);** which reads 9 characters into **message**, or fewer if an end of line character **\n** is encountered, adding the required null at the end. A character other than **\n** can be used as the string-terminator if desired by adding it as a third argument: **cin.get(message , 10 , ',');** will read up to a comma. Incidentally as the terminating character, if found, is not read in, it is still present on the input stream when the next input operation is performed.

6.3.8 Substrings

You often need to access part of a string – either to read from it or to write to it. In **Fortran** this is very easy. If **message** has been declared as a string, then **message(4:8)** refers to the string of characters 4 to 8, and may be used in the same way as any other string,

```
CHARACTER timeofday*30
CHARACTER mess*30 = 'Good morning!'
mess(6:8) = 'eve'
timeofday = mess(6:12)
```

If you omit the first number, as in **message(:4)**, a default of 1 is supplied, if you omit the second, as in **message(6:)**, it defaults to the end of the string.

In **C++** you have to work harder. The start of a string can be altered by incrementing the array: if **alphabet** has the value **0x123456** which is the address of the start of the string, then the second character is at **0x123457** which can be written as **alphabet+1**. Thus **cout << alphabet+3** will print **alphabet** starting with the 4th character. The end of the string is defined by the null character and cannot be altered so easily.

As mentioned earlier, individual characters within a string can readily be accessed and manipulated in **C++**, as they are elements of an array, of type **char** (remembering that arrays start at 0 in **C++**). In **Fortran** they can be accessed as a substring of length 1 but in **C++** they are characters, **'A'**, not strings, **"A"**.

```
alphabet[0]  = 'A';              alphabet(1:1)  = 'A'      ! equally "A"
/* ... */                        ! ...
alphabet[25] = 'Z';              alphabet(26:26) = 'Z'    ! equally "Z"
```

6.3.9   The length of a string

Strings may have different numbers of characters. People's names, for example, vary considerably in length. So do their addresses.

C++ handles this in the fundamental definition of a string. A string is a series of bytes that continues up to and including the string termination character \0. This is also the integer number 0 and the logical value FALSE, which can be useful. If you use pointers then you can reserve space for a string dynamically when you know how large it has to be, though you still have a null at the end.

There is a clear distinction between the *length* of a string, which is the number of characters in the string up to (but not including) a null character, and the *size* of the string, which is the number of bytes reserved for it. The size is fixed at compile time, or by the value given to **new**, but the length may vary between zero and one byte less than the size.

Accordingly, two standard functions are supplied. **strlen(s)** is a specific string function, returning the length of string **s**. If you use it you have to include the header file **string.h**. On the other hand, **sizeof(s)** returns the size of the string **s**. It is a general function, working for any variable: single or array, byte or int or real or whatever. To it, a string is just an array of bytes.

As an experiment, try writing a program as follows, which overwrites the terminator, and see what happens. Treat the exercise as an object lesson both on the dangers of the C++ language, and on its power. We take no responsibility for what happens.

```
//  attempt to change the terminator of a string
#include <iostream.h>
main()
{
    char s[6]  =  "Hello";
    cout  <<  " Final character "  <<  s[5]  <<  endl;
    cout  <<  "Length "  <<  strlen(s)  <<  endl;
    cout  <<  "Size  "  <<  sizeof(s)  <<  endl;
    s[5]  =  '+';
    cout  <<  " Final character "  <<  s[5]  <<  endl;;
    cout  <<  "Length "  <<  strlen(s)  <<  endl;
    cout  <<  "Size  "  <<  sizeof(s)  <<  endl;
    cout  <<  "String is "  <<  s  <<  endl;
    return 0;
}
```

Fortran does not contain variable-length strings. (This is generally reckoned to be a defect, and there is talk of adding them to a future standard.) It goes some way towards attacking this difference: the function **LEN(string)** returns the length of a string, as declared to the compiler, whereas **LEN_TRIM(string)** gives the length of a string, not counting any trailing blanks.

6.3.10   String comparison: equality and sorting

You frequently want to compare two strings to see if they're the same: when looking up an entry in a table, or when comparing a user's request with possible options. In **Fortran** this is easy: the usual relational operators can be used, and you can say **IF(answer == 'YES')**... to check the contents of the string **answer**.

In **C++** it is not so simple: in **if (answer == "YES")**... both **answer** and **"YES"** are evaluated as pointer-to-char, so the expression will evaluate as true only if the string called **answer** is actually pointing to the string **"YES"** given here, and not if it's pointing to another, identical, string somewhere else.

Instead there is a function **strcmp** available in **string.h** which compares two strings and returns 0 (false) if they have no difference (in length or content), so that what is required is **if(strcmp(answer , "YES") == 0)**... A similar function **strncmp(s1 , s2 , n)** is available which does not compare beyond the first **n** characters, and **memcmp(s1 , s2 , n)** performs the comparison of the **n** characters, regardless of any possible termination characters.

If you compare two strings of different lengths in **Fortran**, the shorter one is padded with trailing blanks before the comparison is made. In **C++** two strings of different length cannot be equal to each other. Given the different approaches to the 'length' of a string in the two languages this probably makes sense – but it does mean that **"Hello"** and **"Hello "** will be regarded as equal in **Fortran** and unequal in **C++**.

Sometimes you want to do more than test for equality, you may want to sort strings – say, to produce a list of names in alphabetical order. **strcmp** can be used for this too: what it actually returns is the difference **s1[i]–s2[i]**, where **i** is the point at which the strings start to differ. So that **strcmp("Item B" , "Item A")** evaluates (with the ASCII character set) as $65 - 64 = +1$. In **Fortran**, the operators $>$, $>=$, $<$ and $<=$ can be used to the same effect.

Problem 6.5   *With the above definition, what would be the result of comparing a string of length n with a longer string whose leading n characters match the first, as in* **strcmp("Smith" , "Smithson")**? *Try it out and see. What is the result in* **Fortran***?*

Most of the time this is fine, as the order of letters is sensible. Problems arise if you have to deal with the ordering of characters that aren't letters – does "Item 1" come before or after "Item A"? In ASCII, as it happens, it comes before. With other character sets it may not. So your program may not work on a machine with a different character set. **Fortran** provides a way of comparing strings according to the ASCII character set which works even on non-ASCII systems. These are functions, not operators: **LGE(SA , SB)** is the equivalent of **SA** $>=$ **SB** (or, in old-fashioned Fortran, **SA .GE. SB**) but is guaranteed to be the same on all systems as it uses the ASCII sequence. The logical functions **LGT(strA , strB) , LLE(strA , strB)** and **LLT(strA , strB)** are similarly defined. (The **L** stands for 'lexically'.)

Problem 6.6   *Why is there no* **Fortran** *function* **LEQ***?*

6.3.11   Searching strings

Another useful operation is discovering whether one (short) string occurs anywhere in another (longer) string – for example, you might want to read through a file of text to discover if a particular keyword appears in it. Facilities are provided for this in both languages: in Fortran **INDEX(string , substring)** is an integer function that will return the position of the first appearance of **substring** in **string**, or 0 if no match is found. The equivalent C++ function is **strstr(string , substring)**. This returns the address (within **string**) of the first occurrence of **substring**, or \0 if no match is found.

There is also a function **strchr(s , c)** which matches, if possible, the single character **c** in the string **s**, and **memchr(s , c , n)** is the same except that it scans for **n** characters and doesn't stop if it encounters a null – **Fortran** doesn't need to consider these as separate cases.

Notice the difference between the address (i.e. a pointer) returned by C++ and the location (the character position in the string) returned by **Fortran**.

Problem 6.7   *What outputs would be produced by*
i)   cout << strstr("Hedgehog" , "edge") << endl;   *and*
ii)   PRINT * , INDEX('Hedgehog' , 'edge')?
How could you produce each output in the other language?

In fact **INDEX** has a third, optional, argument of type logical: if it is set TRUE the scanning is done backwards. **strstr** has no such useful feature, but the similar function **strrchr(s , c)** returns the last match rather than the first.

SCAN(string1 , string2) in Fortran returns the location of the first character in **string1** which is also in **string2**. In C++, **strpbrk(string1 , string2)** is equivalent, except that it returns the address rather than the location. **Fortran** provides an optional third argument; **SCAN(string1 , string2 , .TRUE.)** scans backwards to find the last character, as with **INDEX**. If **string2** has only a single character, **SCAN** and **INDEX** are the same.

Sometimes you want to ensure that all the characters in a string you're dealing with are of the type you expect. Again, both languages provide this facility. In **Fortran**, the function **VERIFY(string , set)** returns the position of the first character in **string** that is not in **set**, or zero if all of them are. The C++ equivalent function is **strspn(string , set)** which returns the length of the character string, starting at the beginning of **string**, consisting of characters that appear in **set**. So if you have read a line which should contain only digits, then **VERIFY(line , '0123456789')** had better be zero, and if it isn't it gives the position of the first problem character. **strspn(line , "0123456789")** should be equal to **strlen(line)**, and if it isn't it gives the length of the string *before* the problem – one less than the **Fortran** version.

As before, **Fortran** allows you to work backwards with an optional third argument: **VERIFY(string , set , .TRUE.)** gives the position of the rightmost character not in **set**. C++ does not provide it, however, a similar function **strcspn(string , set)** returns the length of the string, starting at **string[0]**, of characters *not* in **set**.

6.4 AN EXAMPLE PROGRAM

As an example of string and character handling, illustrating many techniques, we present a program to give the molecular weight of any chemical formula. Thus entering **HI** should return **127.9129**, the molecular weight of Hydrogen Iodide. The program must also recognise elements with two-letter symbols like **Si Br**, formulae with numbers such as **H2SO4** (let's not be fussy about subscripts) and formulae with repeated groups involving brackets, such as Ferric Sulphate : **Fe2(SO4)3**.

The program declares arrays for atomic-weight values and for the atomic symbols, which it fills by reading a data file. Then it prompts for the string **formula** and returns the molecular weight given by the function **value_of**. For simplicity we have made the subroutine structure (and the names) the same in both languages, though internal details differ. The **C++** prototypes are given: the formula string is passed as a character pointer. In **Fortran** the variables of the main program are available to all the internal procedures lying between **CONTAINS** and **END PROGRAM**.

```
// program to evaluate Molecular Wts
#include <stdlib.h>
#include <fstream.h>
#include <ctype.h>
#include <string.h>

// function prototypes
  float value_of(char *form);
  float val_head(char *form , int n);
  int   val_count(char *form);
  float val_group(char *form , int n);
  float val_element(char *form, int n);
  void  read_data_file();

// global data
  const int max_elements = 92;
  float AtWt[max_elements+1];
  char symbol[max_elements+1][3];
  int npt = 0;
main ()
{
  char formula[100];
  read_data_file();
  cout << "Please enter formula :";
  cin >> formula;
  cout << "Molecular Weight is "
        << value_of(formula) << endl;
} /* ... function definitions pp 146–150 */
```

```
PROGRAM Molecular_Weight
INTEGER , PARAMETER :: max_el=92
REAL  :: AtWt(max_el+1) , mol_wt
CHARACTER :: formula*100 , symbol&
&(max_el+1)*2 , fmt*11='(A , F15.5)'
INTEGER    :: npt , jstat=0
CALL read_data_file
PRINT * , ' Please enter formula'
READ   * , formula
PRINT * , formula
mol_wt = value_of(formula)
PRINT fmt , ' Molecular Weight is '&
    , mol_wt
STOP ' normal exit to Mol_Weight'
CONTAINS    ! all code in next pages
  SUBROUTINE read_data_file
  REAL RECURSIVE FUNCTION &
    & value_of(form) ! ... etc
  REAL RECURSIVE FUNCTION &
    & val_head(form) ! ... etc
  REAL RECURSIVE FUNCTION &
    & val_element(form) ! ... etc
  INTEGER FUNCTION &
    & val_count(form)
  REAL RECURSIVE FUNCTION &
    & val_group(form) ! ... etc
END PROGRAM Molecular_Weight
```

6.4.1   Reading the data

Suppose there is a file **pt.dat** which contains a list of the chemical symbols (starting in column 1) and their atomic weights, of the form

H	1.0079	Si	28.0855
He	4.0026	S	32.06
Li	6.941	Fe	55.847
C	12.011	Cu	63.546
O	15.9994	I	126.905
F	18.9984	U	238.029

The routine **read_data_file** reads this into the arrays **symbol**, which is an array of strings of 2 characters, and **AtWt**, the corresponding array of real numbers. These must be accessed by the various subprograms, so are declared as global data in the C++ program (see the previous page). The **Fortran** subprograms will be internal to the main program, and thus have access to the data, through **CONTAINS**.

```
void read_data_file()
{
    ifstream datastream("pt.dat");
    while (datastream >> symbol[npt]
        && npt < max_elements)
        datastream >> AtWt[npt++];
    datastream.close();
    cout << npt << " elements read\n";
    return;
}
```

```
SUBROUTINE read_data_file
OPEN(UNIT=1 , FILE='pt.dat')
npt = 1
DO
    READ(1 ,'(A2 , F15.3)' , IOSTAT=&
    &jstat), symbol(npt) , AtWt(npt)
    IF( npt >= max_el .OR. &
    & jstat < 0 ) EXIT
    npt = npt+1
ENDDO
npt = npt−1
PRINT * , ' elements read = ', npt
CLOSE(UNIT=1)
RETURN
END SUBROUTINE read_data_file
```

The data is read one element at a time, and it is checked that there are not too many values for the arrays. The print statement is a reassuring check that the input has worked successfully.

The arrays have 93 elements, one more than there will be data to fill them, unless you include transuranic elements or D for deuterium. This is because the final read statement attempts to fill the next value, and only discovers that it's reached the end of the file afterwards. This could have been avoided, e.g. by reading to intermediate variables and storing them with an updated index only if they have been read successfully, but that gets clumsy in other ways.

If you follow the logic of the changes to the variable **npt**, the number of periodic table elements stored, in both programs, you will see that it **Fortran** there are **npt** in entries 1 to **npt** of the arrays, and in C++ there are **npt** in 0 to **npt−1**.

6.4.2   Analysing the formula string

For our program a chemical formula string, e.g. **Fe2(SO4)3**, consists of a list of items, each of which is either a chemical symbol or a group enclosed in brackets, and which may be followed by a number. (This is not completely general: we cannot handle substances like $CuSO_4.5H_2O$ – unless **CuSO4(H2O)5** is entered.)

To find the molecular weight we thus have to find the weight of the first item, multiply it by the following number if there is one, and then continue with the process, adding the value for the rest of the character string. Eventually we are left with a zero-length string, at which point we stop, the task being over.

This is done by the routine **value_of**. First it checks the length of the string, in case it is zero. If not, it calls two other routines: **val_head** extracts the molecular weight for the first item (which may be a one- or two-letter element, or a group enclosed in brackets) and **val_count** extracts the repeat value, if there is one. As this is also a formula, its weight is found by a further call to **value_of**.

This is a recursive operation: **value_of** is called from within **value_of**. C++ takes this in its stride, whereas in **Fortran** the function must be explicitly declared as recursive. This brings an added complication that the result cannot be referred to by the name of the function and has to be explicitly specified.

In the **C++** version the lengths of the head, the count and the tail are found, and functions called to evaluate them – note the pointer arithmetic of **form+lhead** etc. In **Fortran** the functions nibble away at the start of the string: the **form** returned from **val_head** has had its head removed.

```
float value_ of(char *form)
{
  if(strlen(form) == 0) return 0;
  int  lhead = 0 , lcount = 0;
  if ( form[0] == '(' )
  {
    lhead = 1+strstr(form , ")")—form;
    if (lhead < 0)  cout<< "Missing )!";
  }
  else
  {
    lhead = (islower(form[1]) ?2:1);
  }
  lcount =
    strspn(form+lhead , "1234567890");
  float fhead = val_ head(form , lhead);
  int fcount = val_ count(form+lhead);
  return  fhead*fcount
      +  value_ of(form+lhead+lcount);
}
```

```
REAL RECURSIVE FUNCTION &
& value_ of(form)      RESULT (val)
CHARACTER  :: form*(*)
INTEGER     :: fcount
REAL        :: fhead
IF(LEN_ TRIM(form) == 0) THEN
   val = 0.0
   RETURN
ENDIF

fhead  = val_ head(form)
fcount = val_ count(form)
val = fhead*fcount &
    & + value_ of(form)
RETURN
END FUNCTION value_ of
```

6.4.3   Matching characters in strings

The **val_head** function has the task of decoding the front part of the chemical formula. If the first character is a left bracket, it calls **val_group** to find the matching right bracket (in the **Fortran** version; in the **C++** program it is known to be in the nth element) and extract the value of the group within them, and return what's left of the tail of the formula. If it isn't, the formula presumably starts with a chemical element, so **val_element** is called to pluck off and decode the symbol. The **C++** version is much shorter but, to the unpractised eye at any rate, less immediately comprehensible.

```
float val_head(char* form , int  n)
{
  return(*form == '(' ?
       val_group(form , n)
       : val_element(form , n));
}
```

```
REAL RECURSIVE FUNCTION &
   & val_head(form)    RESULT(val)
CHARACTER :: form*(*)
IF( form(1:1) == '(' ) THEN
   val = val_group(form)
ELSE
   val = val_element(form)
ENDIF
RETURN
END FUNCTION val_head
```

The function **val_element** decodes the symbol at the start of the string and returns the atomic weight. It also has to trim this symbol off the start of the string argument. The symbol may comprise either one or two letters: the second character of the string has to be tested and then included with the first if it is lower case.

```
float val_element(char *form , int n)
{
  int   j;
  for (j = 0; j<npt; j++)
  if (n == strlen(symbol[j])
     && !strncmp(symbol[j] , form , n))
        return AtWt[j];
  cout << "Unknown Element "
       << form << endl;
  exit(EXIT_FAILURE);
  return 0.0;
}
```

```
REAL RECURSIVE FUNCTION &
   & val_element(form)    RESULT(val)
INTEGER      :: i , j
CHARACTER  ::  form*(*) , work*2
i = 1
IF(form(2:2) >= 'a' .AND. &
   form(2:2) <= 'z') i = 2
work = form(1:i)
DO j = 1 , npt
   IF(work == symbol(j)) THEN
      val = AtWt(j)
      form = form(i+1:)
      RETURN
   ENDIF
ENDDO
PRINT * , 'Unknown Element ', form(1:i)
STOP ' ! a serious error found here !'
END FUNCTION val_element
```

6.4.4   Finding digits for a repeat count

val_count picks up possible repeat counts (as in the '2' of 'H2O'). The same computing objective can often be achieved in several different ways. Which way is chosen may depend on many factors, including the personal taste of the programmer. It may also be that different techniques are more appropriate in different languages. This is the case here.

The function is given as its argument the remainder of the formula string. If this string starts with a number (which may have one or more digits) then it has to evaluate it, and also remove it from the front of the formula string. If not, then it should return the default value 1, and the string is unaffected.

```
int val_count(char *form)
{
    int i = 0 , j = 0;
    while(isdigit(form[j]))
        i = 10*i + form[j++] - '0';
    return (i == 0 ? 1 : i);
}
```

```
INTEGER FUNCTION val_count(form)
INTEGER      :: j
CHARACTER :: form*(*)
val_count = 0
DO
    j = INDEX('0123456789' , form(1:1))
    IF (j == 0) THEN
        IF (val_count == 0) val_count=1
        RETURN
    ENDIF
    form = form(2:)
    val_count = 10*val_count + j - 1
ENDDO
END FUNCTION val_count
```

The two routines show two (of many) different approaches to this business of converting a string to a number. In C++ the **isdigit** function (Table 6.1) is used to test the first character of the string (**form[0]**). If it is a digit then its code is converted to the actual value by subtracting the code for zero. This is repeated, multiplying the previous total by 10 at each stage and adding the final digit until a non-digit is found. The function returns this value unless it is zero – this arises if the digit string is of zero length, and means the value returned should be 1.

In **Fortran** the **INDEX** function is used, explicitly giving the possibilities. This serves the dual purpose of ascertaining whether the character is a digit, and if so what its value is: the position in the **'0123456789'** string is 1–10 for the digits 0–9. The **RETURN** is now inside the loop, rather than outside, and the various operations of adjusting the running total and trimming the front of the string each have to be accomplished by a specific statement.

6.4.5   The final piece

The val_group function follows. The first character of the string has been found to be a left bracket, so the matching right bracket has to be found, and then whatever lies between them is itself a chemical formula string which can be evaluated by a recursive call to value_of. Notice how the **Fortran** version copies the group between the brackets to a new string (called **work**), whereas in the **C++** routine **work** is a pointer to the original string, and is terminated with the null symbol.

```
float val_group(char* form , int n)
{
    char *w = new char(n−1);
    strncpy(w , form+1 , n−2);
    w[n−2] = '\0';
    return value_of(w);
}
```

```
REAL RECURSIVE FUNCTION &
& val_group(form)    RESULT(val)
CHARACTER :: form*(*) , work*100
INTEGER  :: i , j
j = 0
DO i = 2 , LEN(form)
  IF( form(i:i) == '(' ) j = j+1
  IF( form(i:i) == ')' ) j = j−1
  IF(j < 0) THEN
      work = form(2:i−1)
      form = form(i+1:)
      val = value_of(work)
      RETURN
  ENDIF
ENDDO
STOP ' Missing right bracket'
END FUNCTION val_group
```

Problem 6.8   *Implement the program. Try it. Try improving it. Use a set of test data which examines the program and its error messages. Such a set might include:*
Li , LiH , HI , SiH4 , UF6 , H2O , H2O2 , Fe2(SO4)3 , CuSO4(H2O)5 , C6H6 , C2H5(OH) , H(CH2)8H , C60
and errors such as LI , Si(H4 , SiH4) , 2HO , Fe(2SO4)3

7

Pointers

First the appearance of pointers at the basic machine-code level is explained. Then their use in C++ is discussed at some length: their declarations and special operators; their connection with arrays and their use with dynamic variables, and some more advanced applications. Finally there is a brief discussion of the more limited pointer facilities of **Fortran**.

Suppose – for some reasons that are here irrelevant – you present someone with a box of chocolates and a seat for a theatre performance. The box you hand over is an actual physical object, but the theatre seat is just a slip of paper with some details – a date, a seat number, and the name of the theatre – printed on it. What you hand over is not the object itself (and gift-wrapping a theatre seat would be rather impractical) but a *pointer* that indicates where the object is to be found.

This use of pointers, as a more compact and convenient alternative to handling the objects themselves, is commonplace in everyday life. Likewise in computing, it is often better for programs not to handle objects directly, particularly if they're large and complicated arrays or other structures, but instead to use pointers to these objects as a quick and convenient alternative.

The ability to use and manipulate pointers is a powerful feature in the original C language and in C++, and so this chapter will devote considerable attention to pointers in this language. In **Fortran**, on the other hand, pointers are a less important feature, and are dealt with in a radically different way, as we shall see.

7.1 POINTERS AT THE MACHINE LEVEL

Pointers appear naturally in machine-code programs, whether generated from Assembler or a higher-level language. At this low level calculations generally fall into three parts:

- Values are loaded from memory into a CPU register or registers.
- The values in the registers are manipulated (by addition, subtraction, etc.).
- The results of the calculation are stored back in memory again.

For example, take the high-level language statement y = x+10. Suppose x and y are single-byte integers stored in memory at locations 1000 and 1001. The assembly language program might look like this. (This is for an 8080 microprocessor. Other microprocessors' assembly languages are different, but not that different.)

```
MOV AX,1000        ; load register AX with x
ADD AX,10          ; add 10 to register AX
MOV 1001,AX        ; store the result in y
```

The only problem is that this is not *relocatable*. These instructions refer to the two memory locations explicitly. When you run a program the operating system loads it into memory; precisely where in memory is not defined beforehand: it can be any convenient chunk of adequate size which is not already in use. You have no guarantee that locations 1000 and 1001 are part of your allocated area. A usable program must be relocatable, with the possibility of running anywhere in memory, so it cannot contain explicit addresses.

To solve this, suppose the register **BX** contains 1000, then the above 3 lines can be replaced by

```
MOV AX,[BX]        ; load register AX with x
ADD AX,10          ; add 10 to register AX
MOV [BX+1],AX      ; store the result in y
```

The square bracket notation such as [BX] produces a machine-code instruction which involves the data at the address held by **BX** rather than the data actually held by **BX**. So if the program starts by loading BX with 1000, or the equivalent address depending on where the program happens to be located in memory (and there are facilities provided to do just this) then all the rest of the program (which in practice will be more than just 3 instructions) will work as it should. The *base register* **BX** contains a pointer to the data, not the data itself.

Thus this manipulation of addresses is fundamental and widespread – indeed it's one of the reasons that digital computers are so powerful. Technical details of assembly language are skipped over here – there are many different indirect and indexed *addressing modes*, all of which use memory locations or registers to point to other memory locations in subtly different ways. It's specially useful for loops over arrays, where a pointer is incremented each time you go through the loop to address array elements in turn. Such features are very well supported by most machine codes. For example, many CPUs can load one register with the contents of a location pointed to by a base register and then increment the base register, all with a single instruction. This means that some apparently involved **C++** expressions translate into very compact code.

So if you could look at an area of a running program used for storing variables, you would find all sorts of different types of data, including variables holding pointers to other variables.

Address	Contents	Variable
1000	1.23	**x**
1004	3.72	**y**
1008	42	**n**
1012 1016	17.71268732176987	**d**
1020	H E L L	**message**
1024	O W O	
1028	R L D 0	
1032	1000	**px**
1036	1008	**pn**
1040	1020	**pmess**
1044	1012	**pd**
1048	1036	**pp**

Fig. 7.1   A typical area of memory in a program

Here you can see some **floats**, an **int**, a **double**, an array of **chars** and some pointers to these variables – and one pointer which points to another pointer. Such pictures helpfully bring out the fact that each variable has (1) an address (2) a value and (3) a name, so as not to get confused between them. Even so, this is a little unrealistic. The compiler knows the name of the variables but it only knows their values when they're initialised; it knows the amount of memory the variable will take up but it doesn't know the absolute address. At run time the address and the value are known, but not the variable name (unless you have a debug flag set). Nevertheless such a picture (often with arrows used to point from the pointer to the address it contains) can be very useful as an aid to understanding.

7.2   POINTERS IN C++

Pointers feature very largely in **C++**. They can be used in powerful ways and many facilities are provided for them. Once you get beyond the simplest level in the **C++** language an involvement with pointers is inevitable.

7.2.1   Declaring pointers

Pointers are declared using an asterisk: the word 'pointer' does not occur in the syntax! You can't just say 'p is a pointer', you have to specify what it's going to point to. **C++** is a strongly typed language, which means that all variables have to be declared before they are used, and the type of the variable must be given as part of the declaration: it's not enough to say 'x is a variable', you have to say whether it is going to be used for real numbers, or integers, or whatever.* In the same way, the nature of the data contained in memory at the contents of any pointer must be specified. For example:

```
float*   pointx;        //  pointer to float
int*     pointj;        //  pointer to integer
shape*   pointsh;       //  pointer to a structure defined earlier
```

Now, although the above declarations make it clear what's going on, it is often better to write them with the * adjacent to the variable, as

```
float    *pointx;       //  pointer to float
int      *pointj;       //  pointer to integer
shape    *pointsh;      //  pointer to a structure defined earlier
```

Although the two forms are completely equivalent, it is good to get into the habit of the latter because if a statement declares more than one pointer, the asterisk must be given for each.

```
float *pointx , *pointy;    //  two pointers to float
float *pointa , pointb;     //  makes pointb a float itself, not  a pointer
```

You can read **int *p;** as 'The contents of **p** are of type **int**' or as '**p** is a pointer-to-int' as appropriate. The statements are equivalent – but one may be a more helpful way of looking at a program, depending on the context.

Pointers-to-pointers are declared in a consistent way. **int **pp;** declares **pp** as a pointer to a (pointer to **int**).

---

* This dictatorial system is actually a tremendous asset, as anyone who's used a loosely typed language (as **Fortran** can be) will acknowledge, as it really cuts down on the potential for introducing bugs.

7.2.2   Pointer operations: * and &

There are two operators specific to pointer manipulation. Both are monadic (single argument) and have high precedence. * – asterisk – is the *contents of* operator and & – ampersand – is the *address of* operator.

'Address of' is applied to a variable (never a constant!) and returns its address; this address is a pointer and can be assigned to a pointer variable, for example

```
int i = 100;
int *p;      // p is declared as pointer—to—int
p = &i;
```

'Contents of' (also known as the *dereferencing operator*) is applied to a pointer and returns the contents of that address. So for the above fragment a subsequent **cout << *p << endl;** would print **100**, and *p can be used in expressions like an ordinary variable: **x = *p + y + 7;**

It can even be used on the left-hand side of an assignment statement:

```
*p = 200;   // sets i to 200
```

This last usage raises two danger flags, one conceptual and the other practical.

The conceptual danger is there are two operations describable as 'assigning to **p**'. The true assignment is **p = &i** – a value (the address of **i**) is placed in the variable **p**. By contrast *p = 200 is an assignment to the *object pointed to* by **p**. It's important to be aware of this distinction and not confuse the container with the thing contained. This is brought out in declarations using **const** in connection with the asterisk:

```
const int *p;          // p points to const int. *p = 200 is illegal
                       // but p = &i and p = &j are OK
int* const p = &i;     // p is unchangeable. Subsequent assignments
                       // of the form p=&... are illegal. But *p = 200 is OK
const int* const p = &i;  // p and *p are unchangeable
```

The practical danger is that even if the pointer **p** has not been assigned a value then *p = 200 is legal (i.e. the compiler will not fault it). At run time the value of 200 will be stuck wherever **p** happens to point to, with random and dangerous results.

Problem 7.1   *If* i *and* j *are of type* int, x *of type float, and pointers* **p** *and* **q** *have been declared as* int *p*; float *q*;, *then which of the following are legal and what do they give?*

i)   i = 8; q = &i; cout << *q;
ii)   p = &j; *p = 9; cout << j;
iii)   q = &x; x = 23.4; cout << 3*(*q);
iv)   i = 8; p = &i; cout << *p;
v)   *p = 9; p = &j; cout << j;
vi)   q = &x; x = 23.4; cout << 3**q;

★ 7.2.3   Further pointer manipulation

Pointers can be assigned to each other, but the compiler is stringent about keeping type integrity pure. Given the declarations **float *r; int *p, *q;** then **p = q;** is legal but **p = r;** is not; the compiler could do it just as easily, but it will refuse to, because it knows that ***p** and ***r** must be interpreted differently so it makes no sense for **p** and **r** to have the same value, and if a programmer thinks they want to do it they're probably making a mistake and introducing a bug. (Yes, there are ways round the smugness of this big-brother compiler. We'll get to them soon.)

You can increment (and decrement) pointers. Having set up **p** to point to the first member of an array, then **p+1** will point to the second member, and **p+1** can be assigned to another pointer, or indeed to **p** itself.

```
int array[100];
int *p, *q;
p = &array[0];                    //  pointer to first int.   p = array is equivalent
q = p+10; cout << *q << endl;  //  print array[10]
```

Incrementing and decrementing pointers may seem trivial but it's actually a very impressive feat of the compiler. Variables of different types occupy different amounts of memory, so when you increment a pointer by one this entails moving it to point to the next byte if it's a pointer-to-char, moving it 4 bytes if it's a pointer-to-int on a machine where the **int** type occupies 32 bits, or appropriate amounts for floating-point numbers of various precision, and user-defined types. The compiler uses the **sizeof** operator and specifies that the pointer be incremented in units of the appropriate amount. It takes care of differences between byte-addressing and word-addressing and the need to start full words on full-word boundaries if your particular system has these features, so you don't have to worry about these technical details.

There are some things you can't do with pointers. You can't add them to each other, and you can't multiply or divide them by anything. You can subtract them, and if they are both pointing to parts of the same array then this gives the difference in position of the two elements; if they're pointing to different objects then the result is meaningless. For example: suppose text is stored in an array **text** and you want to find the length of the first sentence, which will run from the first non-blank character to a full stop. One way to do it is:

```
char *p = &text[0];      // or char *p = text;
while (*p == ' ') p++;
char *q = p;
while (*q != '.') q++;
cout << q - p << endl;
```

Comparisons between two pointers can only be done if they point to the same type (or **void** – §7.2.7.) **p == q** and **p != q** tell you whether **p** and **q** are pointing at the same object. The comparisons **p > q,   p < q,   p <= q** and **p >= q** are legal enough, but only meaningful if **p** and **q** are pointing to the same array.

7.2.4   Pointers and arrays

Pointers and arrays are very closely related in C++ – indeed in many ways they are two ways of looking at the same thing: many expressions can be written equivalently in two ways: in terms of arrays and in terms of pointers.

Suppose an array **a** has been declared as comprising 10 integers (numbered 0 to 9)

int a[10];

then **a[0]** and **a[3]** and **a[i]** all denote locations in memory containing an integer, usable in just the same way as non-array variables declared with **int i , j , k;**.

But there's more to the declaration than that. The elements are stored in memory in order, and array name **a** is a synonym for a pointer to the first element of the array. So the two statements:

i = a[0];
i = *a;

although they look very different, are in fact identical! In the same way, if you want to refer to element 8 of the array, *(a+8) is just as valid as a[8], and &(a[i]) is equivalent to a+i.

As **a** is a pointer, it can be used in pointer expressions:

int *p = a;   // pointer p initialised to start of array
int *q = a+5;   // pointer q initialised to sixth word of array
q[3] = 17;   // set a[8] to 17

This shows how you can effectively set up arrays with arbitrary range; space is allocated for **q[−5]** to **q[4]**. Ex-**Fortran** programmers can index arrays from 1 with statements like **float carray[10] , *farray=carray−1;**.

You can't assign anything to an array name; it is a pointer constant not a pointer variable.

int a[10] , *p;   // declare array a and pointer p
a = p;   // illegal
a += 1;   // illegal

A declaration **int a[10];** does more than a declaration **int *a;**. Both set up **a** as a pointer-to-int. But the first declaration also reserves space for 10 **int**s, initialises **a** to point to that space, and the compiler then knows that the space reserved is the size of 10 **int**s. In the declaration of a function argument no space reservation occurs and the size information is lost, so in argument list specifications (in prototypes and function bodies) they are interchangeable: **fun(int a[ ])** is the same as **fun(int *a)**.

Problem 7.2   *Given* **float*a;** *and* **float b[5];** *what will* **sizeof(a)** *and* **sizeof(b)** *give? What is* **sizeof(a)** *after each of the assignments* **a = &b[0];** *or* **a = b;?**

7.2.5 Pointers and dynamic variables

Programs often have the job of handling an undefined number of objects – for example tracks in a bubble chamber, names and details of personnel, shapes and sizes of electrodes in a field map. To declare an array at the start may be unsatisfactory because you don't know how big to make it – too small and you'll run out of space, too large and your program won't have enough memory to run in.

Instead you can allocate space dynamically as required using the operator **new**. This is applied to a type, which can be intrinsic or user-defined, and it can be an array; it finds where a suitable space is available in memory, and returns a pointer to that space. It also marks such space as in use, so that it won't be overwritten.

> n_electrodes = ... // number of electrodes obtained (only) at run time
> float *xpos = new float[n_electrodes];

Because of the similarity between pointers and arrays, your program can then refer to **xpos[i]** just as if it had been declared as an array.

Dynamic allocation is also often used with large user-defined structures/objects. If you have a large object type **dinosaur** you may declare a pointer to that type in the program (§3.9), and create the actual object only when needed.

```
struct dinosaur
{
    float weight;
    /* ... */
};
dinosaur *p;
p = new dinosaur;
```

Components can be conveniently accessed using the —> member operator: **p—>weight** is a synonym for **(*p).weight**.

Space that's been reserved in this way should be freed again using **delete**, which simply operates on a pointer (which must point to memory allocated by **new**)

> delete [ ] xpos; | delete [ ] dinosaur;

Note those square brackets: they tell **delete** that the associated **new** created an array. They do not actually have to specify the size as that is stored somewhere.

If the space is not available, **new** returns a pointer of zero. (This can be over written – see §7.2.8.) A competent programmer will always check for successful allocation. This can be incorporated in the allocation itself:

```
float* work1 = new float[1000000];      float* work2 = new float[1000000];
if ( work1 )                            if ( !work2 )
    /* ... use the work area */             { cout  <<  "Allocation failed!\n";
else                                          exit(1); }
    cout <<"Allocation failed!\n";       /* ... use the work area */
```

Get into the habit of doing this. It takes a few seconds but can save many hours.

★ 7.2.6   Pointers and multidimensional arrays

Arrays of two or more dimensions can be declared using statements of the form

int a[2][3]

This declares a two-dimensional matrix with six elements divided into two rows of three elements: three columns of two elements.

$$\left(\begin{array}{ccc} a_{00} & a_{01} & a_{02} \\ a_{10} & a_{11} & a_{12} \end{array} \right)$$

The declaration reserves a contiguous block of memory to hold the array – in this case it's six words long. The elements are stored in this block in the order

$$a_{00} \quad a_{01} \quad a_{02} \quad a_{10} \quad a_{11} \quad a_{12}$$

i.e. the last index changes most rapidly. (This is the opposite way round from **Fortran**, in which the first index is the one that changes the most rapidly.) To access element j of row i we just specify a[i][j].

So far so good. But there is another way to store two rows of three numbers (or N rows of M numbers) and that is to construct an array **b** of N pointers each pointing to an array of M values.

int *b[2];
for (int i = 0; i<2; i++) b[i] = new int[3];

With the above two definitions of **a** and **b** the expressions a[i][j], *(a[i]+j), and *(*(a+i)+j) are all legal and all equivalent – remember that q[r] is equivalent to *(q+r) (§7.2.4) – and you can use whichever you choose, even though the first matches the declaration, the others can be used if you want to work in pointer-language; likewise the expressions b[i][j], *(b[i]+j), and *(*(b+i)+j) are all legal and equivalent and you can use whichever you choose, even though the second matches the declaration, the first can be used if you want to work in array-language.

But there is a subtle difference. **b** is a pointer because it is the name of an array. It is a pointer-to-pointer-to-int. Indeed it can be declared as such: it could also have been created with

int **b = new int*[2];
for (int i = 0; i<2; i++) b[i] = new int[3];

When the computer accesses b[i][j] it first gets the value of the pointer-to-pointer-to-int **b**. It increments this by **i**, each increment being the size of a pointer (probably 4 bytes but we don't need to know that) taking it to the ith element of **b**. This is b[i], a pointer-to-int. It takes this value, and points to the start of the appropriate row, that was obtained by the appropriate **new[ ]** invocation when it was initialised. It increments this by **j**, each increment being the size of an **int** (probably 4 bytes but we don't need to know that) so it has a pointer to the desired element.

Now look at the first (array-index) case, **a[i][j]**. **a** is a pointer, and is incremented by **i**. But **a** is a pointer-to-array-of-int, and each increment is worth an entire row of **int** so in this case (and with 4-byte integers) the pointer value to be used increases by, for example, $3 \times 4 = 12$ if **i** is 1. This is now incremented by a further **j** to get to the desired element.

Which method is best? It depends. If you have data in which the 'rows' are of different lengths, then the pointer method is natural. A typical example would be a program that stored the names of several people. Each name would be stored as a string – i.e. an array of **char**, and an array of pointers used to keep track of the separate names.

There is a memory overhead caused by the need to reserve space for the pointer array, but this is small. The pointer method is probably faster. To find the address of an element involves only loading data from memory to registers, and incrementing them by various amounts. By contrast the array method needs a multiplication, which can be slower. However it's unlikely this will affect performance enough to matter.

The pointer method has an overwhelming advantage if the array is to be passed as an argument to a function. With the square-bracket declaration the array dimensions must be passed so that it knows the increment necessary. When you pass such objects as arguments to functions, the interface has to specify whether the object being passed is really a 2D array or a pointer array. Although **(float v[10])** and **(float ∗v)** are equivalent argument lists, **(float m[10][20])** and **(float ∗∗m)** are not! With the square-bracket notation the first dimension size may be omitted. But only the first, not the second, which must be given at compile time. This makes it impossible to use this method to pass an array of arbitrary dimension to a function. The pointer method makes function calls much much easier and more general, at the cost of a slightly more complicated declaration. (If this is a real problem you can easily write a class which will take care of it – classes are described in Chapter 10.)

It is true that with the pointer method there's no guarantee that the data will be in the same area of memory, in contrast to the 2D declaration where the data is contiguous. If you're dealing with large matrices and working on a system with virtual memory, this may give problems in computations with data allocated to very different memory locations, which keeps getting swapped in and out between physical memory and disc. Also, if you're doing repetitive calculations on rows and columns, then a clever compiler can optimise the code by (for example) storing the row offset in a register. If the data has been specified as an array of pointers then this doesn't give it a chance.

The extension to arrays with 3 or more dimensions is straightforward. (Though their thoughtless use can cause you to run out of memory. How much space does the simple declaration **float f[100][100][100][20];** take up? Can you do it on your system?) **f[i] [j] [k] [l]** is handled by the compiler as **∗(∗(∗(∗(f+i)+j)+k)+l)**

Problem 7.3   *Write a function to find the largest element in a general square matrix. Use straightforward array declarations. Then repeat using the pointer-to-pointer method. Which do you find easier?*

★ 7.2.7   Casts

The **C++** compiler is inflexible concerning pointers to objects of different types, so as to save you from blunders. A **char** pointer should point to a sensible **char**. But it also allows would-be experts the freedom to evade these rules by the use of casts(§2.4.10). The cast operator applied to a pointer of one type gives the same address, but with contents treated as a second type; this is dangerous and may be non-portable.

Two notations are possible: with rather clumsy brackets, or function-style. The latter is more elegant but does require a specific **typedef** for use with pointer casts. Here both are applied to unpacking a 4-byte integer (given in hex) into 4 **chars**:

```
int i = 0x4C697645;    // looks like LivE        int i = 0x4C6F4F70;        // 4—byte int
int *p_int = &i;                                 int *p_int = &i;
char *p_c;                                       char *p_c;
p_c = (char *) p_int;    // cast                 typedef char* charpoint;
cout  <<  p_c[0]  <<  p_c[1]  <<                 p_c = charpoint(p_int);
        p_c[2]  <<  p_c[3]  <<  endl;            cout  <<  p_c[0]  <<  p_c[1]  <<
                // but outputs EviL                      p_c[2]  <<  p_c[3]  <<  endl;
```

The situation where such casts are necessary arises if you have a system with different numerical formats. If you have a data acquisition system where one module reads data into RAM in one format, and to get it into a different format used by the rest of the system, you may find yourself manipulating bytes as type **char** and then wishing to use the reformatted 4-byte words as **float**.

Pointers themselves can be cast to integers, and integers can be cast to pointers,

```
int i , j;                    int i = 100;
*p  =  &i;                    *p  =  (int *)i;
j  =  int(p);                 *p  =  12345;      // overwrite word 100
```

This technique can be used to write to specific memory locations, of significance perhaps because particular hardware devices are mapped onto them. This is the **C++** equivalent of the old BASIC **PEEK** and **POKE**. You have to know what you're doing, and random experimentation is not encouraged. Neither is deliberate system corruption! The power that pointers give to a program written by an unscrupulous expert is enormous – which is one reason why **JAVA**, which does not have pointers, is popular for programs which are to be freely ported between machines.

There is also the type of pointer-to-void, which is used by very low-level routines that manipulate data without being concerned what it is, for example to move data between slow and fast memory. There is a bit of a difference in the use of 'void' for functions and pointers. A function **void message(char* c)...** does not return a value. A pointer-to-void **void *p;** can point to a definite real something – it just isn't specified what type that something is. These pointers can freely be assigned to and from (and compared with) pointers to other types, but they can never be dereferenced: given **void *p;** then ***p** can never appear in the program again. They cannot be incremented – the compiler doesn't know how much each increment is worth.

★ 7.2.8   Pointers to functions

C++ pointers can point to functions too. After all, a compiled function is nothing but an area of memory, even though it contains instructions to be executed rather than data to be manipulated. At the machine-code level, to load an address (pointer) and then jump to that address by putting it into the program counter is a perfectly valid and sensible operation.

The compiler continues to be an absolute stickler for complete type declaration. ('A strongly typed language...'). You can't just declare a pointer as pointing to a function: you have to specify the type that the function will return, and the type of each parameter. If you then try and assign this pointer to a function with different return type, or a different number of parameters, or different parameter types, the compiler will not let you. To assign a pointer-to-function you just use the address-of & operator on the name of the function you want it to point to. This can be one of your own functions or a standard library function (provided it's not **inline**).

The syntax contains a necessary but irritating pair of brackets. For example suppose you are doing Fourier transform calculations with either sines or cosines. You could define a function pointer which could be then set to one or the other:

```
double (*trigfun)(double);          //  declare trigfun
trigfun = &sin;                     //  or trigfun = &cos
/* ... */
double y = (*trigfun)(1.24);
```

The differing roles played by the two pairs of brackets in both the first and last lines should be appreciated. The second pair encloses a parameter list, which tells the compiler that what's come just beforehand has to be a function, whether it's being declared, defined, or called. The first pair of brackets are just needed to override the standard order of operator precedence; if you were to write **double *trigfun(double);** and/or **y = *trigfun(1.24);** this would be taken as ***(trigfun(...))**, which is nonsense. Actually **(*trigfun)(1.24)** *can* be rewritten as **trigfun(1.24)**. The compiler sees those brackets, knows they must be preceded by a function, and correctly realises that **trigfun** is a pointer to the function to be called, not the name of the function. The same thing happens when a function is passed as an argument to another function (see §5.2.8). In a call like **float integral = simpson(sin , 0.0 , 1.2 , 50);** invoking **simpson(double f(double) , double xlo , double xhi , int k)** the argument passed as **f** is the pointer to the function **sin**. You might think that the argument should be given as **&sin** and used as ***f** but it isn't. The compiler recognises its nature and takes addresses automatically, just as it does with array arguments.

If you use pointers to functions a lot but you don't like the format in the declaration then the usual way out is to define your own type:

```
typedef float (*fptype)(float);
fptype fpoint;
```

For another example, suppose you're writing a program to be run in several different environments, and it produces messages as it runs. You want to print those messages on the screen, if you're running interactively on a simple system, but append the messages to an output log file of a batch job, or produce voice output if there's a sound card. To cover all these possibilities smoothly you create a function pointer **report** with **void (∗report)(char ∗);** which is used for every such output, **report("Job started");** or **report("Fit not converging: increase tolerance");**, and is assigned (once!) to the appropriate routine by **report = &simplemessage;** or **report = &logmessages;** or whatever.

Stroustrup (1993) advocates the use of arrays of function pointers to provide menus. Suppose you write an interactive program where the user presses a key to select the activity to do next. (**A** for analyse, **E** for edit, **Q** for quit ...) Each activity is done by a specific function, in this case returning **void** and taking no arguments. You can invoke the function corresponding to the appropriate key using an array of pointers to functions, or by a **switch** (as discussed in §4.5).

```
#include <iostream.h>
// Menu using switch
main()
{
    cout << "Type A or Q or E\n";
    char c;
    cin >> c;
    switch(c)
    {
        case 'A' : analyse(); break;
        case 'Q' : quit(); break;
        case 'E' : edit(); break;
        default  : errfun();
    }
}
```

```
#include <iostream.h>
// Menu using array of function pointers
main()
{
    cout << "Type A or Q or E\n";
    char c;
    cin >> c;
    void (∗fp[3])()={&analyse, &quit, &edit};
    char keys[ ] = "AQE";
    char∗ j = strchr(keys , c);
    if(j == 0)
        errfun();
    else
        (∗fp[j−keys])();
}
```

These two programs are about the same length, and the one on the left is (we would say) easier to understand. On the other hand it is easier to expand to more than three options starting from the one on the right.

Another use of pointers-to-functions is to redefine the action taken when **new** fails to find enough space. There is a routine **set_new_handler** in the header file **new.h** which takes as its argument the address of your function

```
#include <iostream.h>
#include <new.h>
#include <stdlib.h>
void newfailure() { cerr << "Not enough memory!!!" << endl; exit(); }
main()
{
    set_new_handler(&newfailure);
    /* ... */
```

★ 7.2.9   Pointers and structures

Pointers to structures/objects are common and useful; a pointer to a class is a specific type. Given **struct vector { float x , y , z; }**; you can declare a pointer-to-vector with **vector *p;**. The member operator —> is useful: **p—>x** is more reader-friendly than **(*p).x** and means the same thing.

In defining class member functions (Chapter 10) there is a handy pointer available called **this** which points to the member being considered. It could be used for accessing members – within a **vector** (above) you can say **this—>x** but such use is superfluous as a long-winded alternative for **x**. It is used if you want to store a pointer to the object: if you are building a complicated structure declared as **vector* struct[100]** as an array of vectors, indexed by **struct_index**, then a routine to add a vector to the structure could be

```
void vector::add_to_struct()
{
    struct[++struct_index] = this;    // should also check for array overflow!
}
```

Pointers to members can be defined using the full scoping syntax: **vector::x *p** declares **p** as a pointer to an *x* component. (But an *x* component is just a float. Consider deeply the difference between pointer-to-float and pointer-to-*x*-component.)

Pointers to member functions get tricky, but this is hidden from the casual user. Suppose you are drawing stars on the screen and a structure (or class – §10.5) **star** contains the position etc, and also four functions to move the star up, down, right and left. In the program you want to take a particular star (or all stars, or all red stars) and move it in some pre-specified way, so you set up a pointer-to-function **move**.

```
struct star
{
    float x , y;
    void move_up(float d) { y += d; }
    void move_left(float d) { x -= d; }
        /* ... */
};
    star pattern[100];          // 100 stars
    void (star::*move)(float);  // move is a pointer—to—member—function—of—star
    move = &star::move_left; // or move_up or...
                             // move all even numbered stars 10 units
    for(int i=0;  i<99;  i+=2) (pattern[i].*move)(10.0F);   // necessary brackets
```

.* connects the pointed-to function with the particular object it's being invoked for. This is a specific operator (with its own entry in Table 2.6). But it reads naturally as: take the contents of the pointer, then apply this to the object. This is a case where remaining in happy ignorance of the subtleties is not such a bad idea. There is an equivalent —>* operator for use when the object is being handled via a pointer.

7.3   POINTERS IN Fortran

The **Fortran** method for pointers is different and more limited. It provides for dynamic allocation of data, but few facilities for pointer manipulation.

Instead of declaring the pointer, you declare an object of the type pointed to, with the *pointer attribute* which means that the declaration does not allocate space for the actual object, but provides the handle to get at such an object which will be *associated* to an actual object later; this association can be done by *allocation* or by *assignment*.

The **ALLOCATE** statement gives space to whatever pointer variable is specified, creating a new object. **DEALLOCATE** frees it again.

```
REAL, POINTER :: p              ! p declared but not allocated
! ... can't use p here...
ALLOCATE(p)                     ! p now allocated
p = 13.7                        ! p is used just like any other REAL
DEALLOCATE(p)                   ! free the memory
! ... can't use p here either ...
```

Notice that the pointer variable is used just like any other variable. There is – in contrast to **C++** – no need for any dereferencing operator as it assumes that when you refer to **p** you intend to refer to the object being pointed to, not the pointer itself.

Allocating and deallocating a single variable isn't particularly useful. You're more likely to need this when you're handling large arrays or user-defined structures which take up a non-negligible amount of your computer's memory. This is easily done as the dimensions of an array declared as a pointer are not specified at declaration time (though their number is).

```
REAL , DIMENSION(: , : , :) , POINTER  :: cube
! ...
N = 50
ALLOCATE(cube(N , N , N))
! ...
DEALLOCATE(cube)
```

Assignment associates a pointer to an already existing object (as opposed to allocation, which associates it to a new one.) Because the usual expression syntax assumes that we want to refer to the object pointed at rather than the pointer itself, a special operator is needed for this. This is the *pointer assignment operator* =>.

```
REAL , TARGET    :: x , y
REAL , POINTER :: p , q
  ! ...
y = 1.234
p => x                        ! pointer assign p to x
p = y                         ! normal assignment of y to p
PRINT * , x                   ! will give 1.234
q => p                        ! x, p, and q are now equivalent
```

Note the contrast between **p** => **x** and **p** = **y**, and the difference between the two pointer assignments in this fragment. In the first assignment the pointer **p** is associated to the variable **x**. In the second assignment the pointer **q** is allocated to the same object as is the pointer **p**.

Problem 7.4   *If the pointer **p** is then reassigned with, say, **p** => **y**, what does **q** now point to?*

Notice the **TARGET** attribute in the declaration of **x** and **y**. **Fortran** is very concerned with the optimisation of compiled code. This is a lot more difficult if the compiler doesn't know which variables are being referred to: it may look as if an array **aaa** is not used in a section of code, as the name **aaa** doesn't appear in the program anywhere, which is mostly concerned with doing complicated things to the pointer **ppp**. In fact the pointer **ppp** is assigned to **aaa** – but this was done in another piece of code and you can't expect the compiler to know that! Therefore the rule is imposed that pointers can only be assigned to other variables if those variables have the **TARGET** attribute; this is a sign to the compiler that they may be accessed in non-obvious ways, and it should not be too clever in optimising their use. (Pointers can be allocated to other pointers without the **TARGET** attribute being specified – if an object can legally be pointed to by the first pointer, it can be pointed to by the second.)

Types must match in pointer assignments. Pointers declared as real can only be assigned to reals, integers to integers, and so on. If arrays are concerned then the ranks must match. This can sometimes be used as a concise way of accessing a particular sub-array.

```
REAL , TARGET    :: allgrids(10 , 10 , 1000)
REAL , POINTER :: thisgrid(: , :)
  ! ...
thisgrid => allgrids(; , ; , 15)
  ! ...
NULLIFY(thisgrid)
```

Generally when you've finished with a pointer you don't care what happens to it, but occasionally you might worry lest a program error caused it to be used later. You can safeguard against this by **NULLIFY** to disassociate the pointer, and it is probably good practice.

8

Input and Output

The ways of formatting output are given for **Fortran** and the two alternative I/O systems of C++, with a briefer discussion of input. Various peculiar (and useful) features of the various systems are pointed out. An example program is given to compare the three. Then there is an account of the mechanics of reading data from files and writing data to them, which is essential for serious data processing.

Reading data from the keyboard and displaying it on the screen is easy and straightforward in either language,

`cin >> x;` `cout << "You typed " << x << endl;`	`READ *,x` `PRINT *,'You typed ',x`

Such simple input and output operations are all you need to begin with. But soon you will want more than this. Sometimes you will want to specify the exact *format* of what is printed (or, less frequently, what is read): for example when you want a value printed to some specific number of decimal places, or to print several numbers lined up in a neat column. Sometimes you may need to use devices other than the keyboard and screen, such as files of data on disc.

8.1   C++: STANDARD I/O AND STREAM I/O

Within **C++** there are two different systems for doing input and output: *standard* I/O and *stream* I/O. Standard I/O is the system used in the original **C** language. When Stroustrup produced **C++** he developed a new I/O system as part of the new language, but did not remove the possibility of using the previous system.

The stream I/O system is closer to the 'spirit' of **C++**, and is advocated by **C++** purists and the object-oriented programming lobby. It has the great advantage that you can write I/O routines for object types that you define yourself. But there are many people who say the standard system is easier to use, and that the stream I/O system is unnecessarily complicated. The rest of this chapter will consider both systems and bring out the differences, so you can decide for yourself.

For consistency, and as a personal preference, we have used the stream I/O methods of **C++** throughout the rest of this book. But it's worth knowing about both systems; if you only know about one, you may well one day find yourself in a programming environment with people who only use the other.

8.1.1   Standard I/O in C and C++

To use standard I/O, include the header file **stdio.h**, instead of **iostream.h** for stream I/O. If you want to use both – and this is possible, though it's not a particularly good idea – then you include both. (You also have to call a function **ios::sync_with_stdio( )** to ensure the outputs from the two systems keep in step.)

The workhorse of the standard I/O system is the output statement

```
printf(format–string , arguments ...);
```

which prints the list of arguments onto the standard output stream (usually the screen) according to the format string (to be described in the next section). There is a corresponding **scanf** statement which reads from standard input (usually the keyboard). Because this is a function which affects its argument(s) they have to be passed as addresses (**C** does not have call-by-reference).

```
scanf(format–string , argument addresses ...);
```

The **scanf** and **printf** functions return the number of items that have been read or written as appropriate. This is not particularly useful – though careful programmers make use of the fact that if **printf** fails for some reason, it returns a negative value, and if **scanf** gets to the end of its input file it returns the constant **EOF** (in **stdio.h**).

Here are some simple examples – the details follow in the next section:

```
printf(" Answer is %8.3f centimetres \n", x);
printf(" Processing starts");              // the argument list can be null
scanf("%f", &tmp);                         // fetch a real into tmp
puts("Please give start temperature ");  scanf("%f", &temp);
gets(text_string);    // puts() , gets() are used to display and read strings
```

8.2   FORMATTING OUTPUT IN Fortran AND C++/C STANDARD I/O

When a variable or an expression is sent to the output stream its value has to be *formatted* as a string of characters. For example the value '1' must be turned into the ASCII character for 1, which is 49. The value 10 becomes two bytes: 49 and 48. A floating-point number such as '1.2' needs three bytes: 49 for '1', then 46 for '.', then 50 for '2'.

There are three things that the compiler needs to know and you may have to specify as you get beyond the **PRINT** * , **x** level of sophistication:

• Extra material. You may want to add words like 'The answer is ' or 'Value after first iteration', and perhaps spaces before/after the printed number.

• The type of the variable. Are these 32 bits to be interpreted as a real or an integer or four ASCII characters or what?

• The details of how to print it: base, style, width and precision. Do you want to print **12.34** or **12.34000** or **0.1234 E02** or **12.3**? Do you want it to start at some point across the page? Maybe you want to print **255** as **FF** in hex or **377** in octal.

8.2.1   The format string and the I/O list

The **C++** stream I/O system deals automatically with the first two points, and so you need only concern yourself with the third. **C++/C** standard I/O and **Fortran** systems deal with all three. These last two systems are actually rather similar, so let's look at them together. In both cases the instructions are encapsulated in a *format string* and an *I/O list*.

printf(formatstring , list);	|	PRINT formatstring , list

For example

printf("Goodbye World \n");	PRINT '(" Green world")'
printf("Answer %8.3f \n" , x);	PRINT '(" Answer " , F6.3)', x
printf("Answers %4i %9.4f %6.3f \n",	PRINT '(" Answers " ,I4 ,F9.4 ,F8.3)'&
j , x , y);	& , j , x , y
printf("For %4i hits \n Q = %9.2f\n",	PRINT '(" For" , I4 , " hits " , / ,&
n , Q);	& " Q = " , F9.2)' , n , Q

Notice that the brackets in **Fortran** are a mandatory part of the format string, whereas in **C++** they are part of the **printf** function syntax, so the I/O list comes outside the brackets in **Fortran** but inside them in **C++**.

Since the format strings *are* strings they can be stored in string variables, often with a noticeable gain in clarity. (You can also vary formats with string manipulations.)

char fmt1[ ] =	CHARACTER :: fmt1*30
"Values %4i %9.4f %6.3f \n";	fmt1 = '(" Values " , I4 , F9.4 , F8.3)'
printf(fmt1 , j , x , y);	PRINT fmt1 , j , x , y

8.2.2   Details of the format specifications

In **C++/C** standard I/O the format specification is best thought of as a string containing

- *ordinary characters* which are just sent to the output,
- *data descriptors* introduced by the **%** character, which are replaced by items from the I/O list,
- *escape sequence* characters such as **\n**, for which the appropriate character is substituted and sent to the output. See the full list in §6.1.4.

In **Fortran** it is better to think of the format as specifying a list of items, separated (usually) by commas. These items fall into 3 types, corresponding to the same purposes: they can be

- *text strings*, which go straight to the output,
- *data descriptors*, which produce the output of the next item from the argument list,
- *control descriptors*, which may produce blanks or new pages, or may affect the format of subsequent data items. Details appear in §8.2.3 with some more in §6.1.4.

There are great similarities between the **Fortran** and **C++/C** data descriptors: both use letters to decide the general way the item is to be printed, and numbers to give details. In **C++/C** they are introduced by a percentage symbol, **%**, to make clear they are not part of the text: in **Fortran** there is no need of such a symbol as the descriptors appear separated by commas.

Table 8.1   Format specifiers for controlling both input and output streams

To specify	Fortran	C++	Notes
Integer	**I**	**i** or **d**	**i** ('integer') and **d** ('decimal') are the same on output
Binary	**B**	None	C programmers all read hex and octal fluently anyway
Octal	**O**	**o**	That is 'O' or 'o' for 'octal', not zero
Hex	**Z**	**x** or **X**	**x** and **X** print digits as **a..f** and **A..F** respectively
Floating point	**F**	**f**	Fixed point representation e.g. 123.45
Floating point	**E**	**e** or **E**	Exponent representation e.g. 1.2345E2
Floating point	**G**	**g** or **G**	Chooses fixed point or exponent according to size
Unsigned integer	None	**u**	Leftmost bit counts $+2^{n-1}$ not -2^{n-1}
Character	None	**c**	Print integer as character – not applicable in **F90**
String	**A**	**s**	
Logical	**L**	None	Prints **T** or **F** (The **C++ bool** will also provide this)

One can specify the *width* **w** of an item and (sometimes) the *precision* **p**. (The width **w** is optional in **C++/C** but necessary in **Fortran** except for the **A** format. The precision **p** is optional in **C++/C** and for **Fortran I, O, B, Z**: it cannot be given with **L**.) The width **w** includes all signs, decimal points, etc. If the natural width of the item is less than this, it is padded with blanks (usually at the front – see later). **p** is the number of digits printed after the decimal point. **w** and **p** are specified similarly in the two languages as two numbers separated by a full stop, the difference being that these come after the letter in **Fortran** (**F8.4**) and before it in **C++/C** (**%8.4f**). If **p** is omitted the full stop is omitted too: **I8** or **%8i** as appropriate.

If the number you try to print is too big for the width specified, then the actions taken are different: in **Fortran** a field of the specified width is printed, but filled with asterisks. In **C++/C** the width is expanded till it's big enough for the number you've given.

To include new lines in **C++/C** you include the newline character \n in the format string. This is commonly done at the end of the string, but they can also be included anywhere else if that's what you want. **Fortran** provides a new line automatically at the end of the print statement: to move to a new line in the midst of an output statement, a slash / is inserted as a control descriptor in the list.

Problem 8.1   *If* **x** *has the value 12.345 and* **k** *has the value 45678, what would be produced by*

i)   **PRINT '(" " , I6)' , k**
ii)   **PRINT '(" " , F6.2)' , x**
iii)   **printf(" % 6i\n" , k);**
iv)   **printf(" % 6.2f\n" , x);**
v)   **PRINT '(" " , F6.2)' , −x**
vi)   **printf("%d %i %o %x \n" , k , k , k , k);**
vii)   **PRINT '(" " , F6.2 , / , " " , I6)' , x , k**

★ 8.2.3   More detailed features

In **C++/C** various *flags* can be inserted between the percentage sign and the width. These are [+ − space # 0] and have the following effects.

+ as in **%+8i**   All numbers are preceded by their sign + −.

as in **% 8i**   The + sign is replaced by a space.

− as in **%−8i**   Output is left-justified, 1234b̷b̷b̷b̷ rather than b̷b̷b̷b̷1234. The symbol b̷ means a blank character. This format is useful for strings.

0 as in **%08i**   The number is left-padded with zeros.

is the *alternative* format in which the **e, E, F, g, G** formats always contain the decimal point, and the **g, G** formats retain trailing zeros. In addition the **o, x, X** formats start with **0, 0x, 0X** as appropriate.

In **Fortran**, to get + − signs include **SP** (sign print) as an item in the format statement (but this affects all subsequent numbers printed by this statement: you have to use **SS** − sign suppress − to switch the effect off again).

In **C++/C** the items in the list must match the data descriptors in the format string. There must be the same number of each and the types must be correct. Excess items are not printed but unmatched descriptors produce rubbish. (The **printf** and **scanf** functions take a variable number of arguments, which is cleverly done. The functions use the format string to discover how many more arguments are to be passed.)

Fortran is more flexible: if there are more list items than descriptors then a new record is started, and the format string is recycled from the beginning of the last parentheses; if there are fewer then the items in the format string are output until there is a format data descriptor with no matching list item.

Problem 8.2   *If* i, j, *and* k *have values 101, 102, and 103, what is wrong with* **PRINT '(" " , I3 , I3 , I3)'** , i , j , k?

Problem 8.3   *What happens on your system with* **printf(" %i4 "**, x); *or* **PRINT '(I4)'** , x, *where* x *is real? (This is undefined in the language specification.)*

Real numbers, such as $-9.876 \cdot 10^{-2}$, can be output in two ways. With fixed-point form, using **Fw.p** or **%w.pf** in the two languages, you get **p** figures after the decimal point and **w–p–1** spaces before it for leading digits and (if necessary) a sign. In exponential form, using **E** or **e**, you get a number-value of precision **p** significant figures (a *significand*) with an exponent, all in a total width **w** – which should be safely larger than **p** to allow space for the exponent, minus signs, etc. The **G** or **g** format switches between fixed point for numbers that are not too large or too small and exponential form for large or small numbers where it makes more sense, and its use is recommended. In C++ the **E** and **G** formats produce an upper-case **E** in the output, **–9.876 E–02**, and the lower-case **e** and **g** give **–9.876 e–02**.

Fortran has the extra forms **EN** and **ES**. The 3 possibilities are **E** for exponential (**–0.98760E–03** with a leading 0. or .), **EN** for engineering (**–98.760E–03** where the exponent is a power of 3 – Mega, Kilo, milli, micro, nano, pico, etc – and the significand lies between 1. and 1000.), and **ES** for scientific (**–9.8760E–02** in which the significand lies between 1. and 10.). The number of digits given in the exponent can be specified by the optional **E** value: **E15.5E4** allocates 4 digits to the printed exponent and prints **–0.988E–0003**.

Character values use **A** or **Aw** in **Fortran**. If **w** is not specified then the length of the string is taken. If it is specified then the string is truncated from the right or padded at the left as necessary. Thus if 'VALUE' is output with the format descriptor **A** it produces VALUE, with **A4** then VALU, and with **A8** then ᵇᵇᵇVALUE is printed. If a width **w** and precision **p** are provided with the C++/C equivalent **%s** then the string (up to the null string terminator) is printed, truncating to **p** characters if necessary, and padding at the left to give a total of **w** characters.

C++/C uses **%s** for strings and **%c** for single characters. It also provides two functions for single character I/O which are sometimes more elegant than the equivalent **printf("%c"** , c) and **scanf("%c"** , &c). **putchar(char c)** outputs the single character c. Its converse is **getchar()** (no arguments) which returns a character – essentially what you get when you press a key on the keyboard. This is particularly useful for things like menu programs. **getchar** returns the constant **EOF** if the file being read comes to an end. (The return value of **getchar** is actually an **int**, not **char**, as **EOF** – whatever it is – is not in the range of possible character values.)

★ 8.2.4   Various other Fortran FORMAT features

★ 8.2.4.1   PRINT and WRITE

The **PRINT** statement is the simpler way of controlling formatted output. We have used it extensively.

PRINT * , start , finish	! use default format
PRINT '(2I10)' , start , finish	! format is specified in the string
PRINT 100 , start , finish	! format is specified at label 100

The more general output statement is **WRITE**, which is similar to **PRINT** except that it also involves a unit-number (§8.7). A **PRINT** statement is essentially the same as a **WRITE** statement to standard output (usually unit 6).

WRITE (6 , *) start , finish	! use default format
WRITE (6 , '(2I10)') start , finish	! format is specified in the string
WRITE (6 , 100) start , finish	! format is specified at label 100

The **WRITE** statement has two optional parameters, **IOSTAT, ERR** which can be used to recover from an output problem. If an output is unsuccessful – because the printer is jammed, or the disc is full, or whatever – this will (probably) not crash the program, so you have a chance to catch the problem and recover in some way. The I/O status integer variable set as **IOSTAT** has the value 0 for a successful write, a positive value if an error was detected, and a negative value if an *end-of-file* was encountered, so you can look at that.

```
WRITE (14 , fmt ,  IOSTAT=ios) x , y , z
IF( ios /= 0 ) CALL WRITE_ERROR(ios)  ! deal with the error in some way
```

The **ERR** parameter is a label number, and control passes to it if something goes wrong. Although labels are generally a bad thing, as they can lead to unreadable code, this is one instance where their careful use can help rather than harm.

```
      WRITE(14 , fmt , ERR=999)  x , y , z
      RETURN
999   PRINT * , ' Output Error'
      STOP
```

If you have many output statements you can handle them all with one error label. This is very little extra work; lazy programmers who dislike the hassle of adding a test to the I/O status variable after every output statement may not bother, but they usually manage to add the few extra characters needed to specify an error label.

There is no separate statement needed if you want to read from a file: you use the usual **READ** but with a unit number specified.

```
READ * , x    ! standard input       | READ(12 , * , ERR=999) x    ! from unit 12
```

★ *8.2.4.2   The stand-alone* **FORMAT** *statement*

Another feature of **Fortran** is the use of the separate **FORMAT** statement. These two are equivalent:

```
PRINT '(" The answer is " , F6.3)' , x  |      PRINT 100 , x
                                         |  100  FORMAT(' The answer is ' , F6.3)
```

The stand-alone **FORMAT** statement is useful:

1: If you have several I/O statements using the same **FORMAT**, then by specifying the **FORMAT** separately you need only do so once.

2: If the **FORMAT** is long and complicated, especially if it involves quotation marks, it may help to separate it from the rest of the I/O statement.

Its disadvantage is that the **FORMAT** statement can come anywhere in the entire program, and may be several pages away from the line that uses it. This is a special pain given the habit some programmers have of putting all their **FORMAT** statements together at the end of the program. This results in code like

```
IF (condition) THEN
    PRINT 100
ELSE
    PRINT 101
ENDIF
```

which is incomprehensible to outsiders (and insiders too, probably). On balance, we would view the labelled **FORMAT** statement as something to be avoided except in odd and occasional circumstances.

★ *8.2.4.3   Default* **FORMAT**

Fortran has the friendly feature that the whole **FORMAT** statement can be replaced by a single asterisk. It then adopts default values for bases, widths, and all the rest. And quite honestly, most of the time this is pretty satisfactory. It's only when you're working on something that you want to look exactly right to impress your boss or the customers, or when you want a table of numbers of different sizes to line up properly, that you need to specify the **FORMAT** explicitly.

★ *8.2.4.4   Repeat counts*

When an item in a **FORMAT** statement is repeated, this can be specified by giving a repeat count in front. More complex sets of items can be repeated using brackets. Thus **I6 , I6** can be replaced by **2I6** and **3(' X = ' , I3 , F7.3)** is equivalent to the sequence **' X = ' , I3 , F7.3 , ' X = ' , I3 , F7.3 , ' X = ' , I3 , F7.3**.

★ *8.2.4.5   Carriage-control characters*

In **Fortran** (the dead hand of history strikes again) the first character in a line has a special significance and is often not printed. Or in some systems it is printed normally on the terminal but has special effects on a printer – for example a '1' generates a new page. It is usually wisest to make sure the first character is a blank. You may have noticed in some of the previous sections that the **Fortran** examples had a single character space **" "** or **' '** at the start, and this is why. It's actually helpful: given that you need to print a text string anyway, you might as well use a few more characters and add some text to make the output meaningful. (A blank is also produced by **1X**.)

★ *8.2.4.6   Loops in I/O lists*

A **Fortran** I/O list can specify several items, typically elements of an array, using an *implied loop*. For example

> **WRITE (6 , '(8I5)')**   (ind(j) , j=1,8)

will print the eight numbers **ind(1)** to **ind(8)**. The index variable (here **j**) is a standard variable, so be careful that it's not one you're using for another purpose. More complicated loops are possible too:

> **WRITE (6 , '(" " , 5F8.3)')** ((array(j , k) , k=1,5) , j=1,3)
> **WRITE (6 , '(1X , I4 , F8.3 , " +− " , F6.3)')** (j , val(j) , err(j) , j=1,n)

Notice how each implied loop is demarcated by a set of brackets.

This example uses the way **Fortran** behaves when the output list is longer than the format list. It outputs a space followed by the 5 columns of the 1st row: **array(1 , 1)** to **array(1 , 5)** in **F8.3** format. The I/O list continues with **array(2 , 1)**: the **FORMAT** has run out of items so it ends the record (i.e. starts a new line) and starts at the beginning again with another space, and 5 more numbers. It has to do this yet again, for the third row, before the output is finally complete.

Problem 8.4   *How could you use a* **WRITE** *statement to print a matrix with the row number given at the start of each row?*

If you want to print an entire array then you don't actually need these loops, you can just use either of

> **PRINT ∗ , array** | **PRINT ∗ , TRANSPOSE(array)**

and you get all the elements. However these are just printed in column order (§3.10.2) which may be confusing for an array with more than 1 dimension. If you want a row format you can put in the loops yourself, or, more easily, output the transpose (see §9.8), the columns of which are the original rows.

★8.3   FORMATTED INPUT IN Fortran AND C++/C STANDARD I/O

Input is similar to output in both systems: a list of variables to be input is matched to a format string.

For **Fortran** input the default ✦ format is sensible and it understands what it is reading. In **C++/C** you can use **%i** for integers and **%f** for reals, and they will also be read with due attention to decimal points, spaces, etc. When you are reading a value from the keyboard this is almost certainly all you need.

You may need more control when your program reads a file produced by another program, with lots of numbers in particular formats. Then you can specify the width and precision, even though it is unlikely that you care whether a number you read has the format 1.234 or 1.234000 or 0.1234E1. For floating-point numbers the language is sensibly flexible: the decimal point on the input overrides the specified value of **p**. Not so with the width **w**: if the number **12.345** is read with **F5.1** it becomes **12.34**. Integers are worse: **1234** read with **I5** appears as **12340** (spaces are treated as zeros). You have to be absolutely certain of the field width of the input number, e.g.

> scanf("%4i" , &j); | READ   '(I4)' , J

reads an integer which had better be 4 digits long.

For **scanf** the format items are the same as in **printf** except **%i** reads integers in hexadecimal (0x1234), decimal (1234), or octal (01234), whereas **%d** reads in decimal. On output **%i**, **%d** give signed decimal integers.

It may be that you need to read numerical values from standard input (**scanf( )**) or a file (using **fscanf( )**, §8.7) which also contains text: if the input is **78 Entries in table** or **Density correction factor 1.06** your program wants to read **78** or **1.06** ignoring the helpful messages. The first can be read as **scanf("%d" , &entry)**. More generally this can be done using the *square-bracket format*. **%[abc]** is equivalent to **%s** except that only the characters **a**, **b**, or **c** are read – anything else terminates the string and input moves on to the next item. Thus

> char s[100];
> scanf("%[1234567890]" , s);

will read characters into the string until it encounters something which is not a digit. That works for the '78 Entries' example, except '78' is read in as a character string rather than the integer you need – we'll come back to that in a moment.

For the second case above you can use **% [ˆabc]** which reads characters up until it meets one *not* amongst those given

> char s[100];
> double dfactor;
> scanf("%[ˆ1234567890]" , s);   // read up to (not including) 1st digit
> scanf("%f" , &dfactor);   // read the wanted value

Problem 8.5   *In the above two* **scanf** *statements, why is there an ampersand in the second but not the first?*

When you need to read data from a string this can be done using the function **sscanf** which acts similarly to **scanf** but takes input from a string, given as the first argument. So to decode that '78', as promised above, you say

```
int entries;
sscanf(s , "%i" , &entries);
```

There is a matching output function **sprintf** which stores formatted output in a string.

In **Fortran** these facilities are neatly provided for: if you give the name of a character array instead of a number as the unit in a **READ** (or **WRITE**) statement then the input (or output) item list comes from (or goes to) that array, according to the specified format.

```
! decode a line of data of the form   text = value
      CHARACTER :: line*80
      READ (5 , *)  line                  ! or READ *,line
      i = INDEX(line , '=')               ! find equals sign
      READ(line(i+1:) , '(F10.3)') val    ! read value from array
```

Problem 8.6   *A string contains a name and address, separated by a slash – as in* **John Smith/10 Mornington Crescent/Manchester.** *Write a program to print this on separate lines. Try using (a)* **sscanf** *in* **C++.** *(b) Reading from an array in* **Fortran.** *(c) String manipulation in either language.*

It is very easy to ignore input items in **Fortran** if you know how many characters they hold: the format descriptor **X** will miss a character (on output it prints a space) and this can be used with a repeat count. So

```
      READ  '(10X , I2)' , n
```

will skip 10 characters (which could be helpful text) and then read a 2-digit integer.

In **C++**, unlike **Fortran**, the width specifications can be omitted and generally are – except when reading strings, as they specify a maximum string length and can save you from overflows if the input string is longer than the space allocated to the variable.

You do sometimes meet a problem when inputting strings containing spaces.

```
      CHARACTER :: name*20
      READ * , name
```

will work if you give it **Fred** or **Smith** but not **Fred Smith** or **Smith, Fred** as it will take these as two separate items. To get round this you have to put the whole string you want to handle within quotes. In **C++/C** this doesn't help – the **%s** edit descriptor will read quotes like any other character. You can use the square-bracket format or read the line as string of bytes and unpack it with **sscanf** or else use **gets**.

Problem 8.7   *Write a program to read a line of data (i.e. up to* \n*) into a string using* **getchar.**

8.4   C++ STREAM I/O

We have used statements like

cout << a << b << c << endl;

to print several items (3 in this case) on the screen. Now let's explore more carefully the syntax of this statement.

First look at **cout**. This is an object with a particular type: **ostream**. If you read the header file **iostream.h** which you need to include in every program, you'll find it declared there, and you'll also notice that there are two other objects **cerr** and **clog** that also have this type. You can use these – try them! Although they all come up on the screen when you run normally, it can be very handy when you're running a long job, perhaps in batch mode, to be able to redirect the error messages and log messages to separate streams, so that, for example, important warnings don't get lost in hundreds of lines of numerical output. **clog** is buffered, so large amounts of data can be output efficiently, whereas **cerr** is not, as you want to get any error messages as soon as they occur. Each stream is connected to an output device – the screen, or a file, or a printer, or a pipe, or whatever.

Now consider the double arrow << which is actually an operator (sometimes known as *put-to*, sometimes as *left-chevron*). Normally it's the shift operator, used for manipulating binary numbers, but it has been overloaded: in the expression **a << b** then if the left operand **a** has type **ostream** – such as **cout** and **cerr** and similar – then it does not try to shift **a** up **b** binary places, instead it takes the quantity **b**, produces from it a set of characters in a way which depends on the type of **b** and other conditions that we'll come to in a moment, and inserts these characters in the stream **a**. It also returns a value which is just a reference to the stream **a**.

So what happens with the expression **cout << x << y** is that* this is evaluated as **(cout << x) << y**. The first << operator prints **x**, and the second now finds itself with a value on the left which is just the stream **cout**, so it prints the quantity – here **y** – to its right.

Why bother to know all this? Well, for one reason it shows you why **cout << x*y;** will work, but **cout << x&y;** will not. The << operator has higher precedence than **&**, so the compiler takes it as **(cout << x)&y;**. Brackets are needed to get round the problem.

The object **cin** is defined in **iostream.h** as being type **istream**. This is very similar to **ostream**, with the other shift operator, >> (sometimes known as *get-from*, sometimes as *right-chevron*) overloaded to mean that **cin >> a** extracts a stream of bytes from **cin** and forms a value which is stored in **a**. The value of the expression **cin >> a** is **cin** so again several input items can be chained: **cin >> a >> b** reads **a** and then **b**. Unlike **scanf** we don't need messy ampersands!

---

* << is, like most operators, left-associative.

8.4.1   Formatting output with C++ stream I/O

Why are there so few formatting facilities in **C++** stream I/O? Because more are not necessary. Think again about the first point of §8.2, the inclusion of extra messages ('literal characters'). As the items in the I/O list can actually be character strings, the edit-string facility provides two ways of printing a character string: as part of the I/O list or as part of the format.

printf("%s" , "Farewell, you");	**PRINT** '(A)' , ' Cruel World'
printf(" Big, bad, World\n");	**PRINT** '(" One World")'

The **C++** stream I/O removes this redundancy. If you want to send a character string to the output stream, then you do so in its own right.

> cout << "Global World";

The second point of §8.2 was the need to tell the compiler what type of variable is being output. Stream I/O avoids this on the grounds that the compiler knows this perfectly well already. The bugbear of the standard I/O method is that the format items and the I/O list must agree: there must be the same number of each, and each list item must have a format item of the correct type. Not having to do this matching with stream I/O is a real liberation.

You still may need to control the widths of printed numbers to make them line up, for example. This control is achieved by adding *manipulators* to the output stream. These objects are not printed, but affect subsequent printing. They are defined in the header file **iomanip.h** so you have to include this, in addition to **iostream.h**. The width and precision, the equivalents of **w.p** in **printf** or **Fortran**, are the manipulators **setw(w)** and **setprecision(p)**. To print **x** with a precision of 4 and a width of 8 you say

> cout << setw(8) << setprecision(4) << x << endl;

This may seem complicated but it gets worse. If you have very large or very small numbers and want to print them in 'scientific' notation, with an exponent, then this requires the manipulator **setiosflags(ios::scientific)**. This additional complication is one example of the ability to change the *I/O state flags*, which are a set of bits used to control details of the way input and output are performed with this stream. If the **ios::scientific** flag is set then all numbers will be printed as **1.234 e1** etc until you send the manipulator **resetiosflags(ios::scientific)** to the output stream.

There is a difference between the **ios** flags and other manipulators. When set they stay set, whereas **setw** and **setprecision** only affect the next item in the list, not even subsequent items in the same statement.

Members of the **ios** class are given with the double colon (**ios::scientific** etc) – and it also means that they don't get confused with your own variables which may have the same names.

The **C++** stream I/O manipulators are shown in Table 8.2.

Table 8.2   C++ stream I/O formatting

Manipulator	Meaning	Fortran analogue	C++/C analogue
setw(w)	width	w in **Iw** etc	w in **%wi** etc
setprecision(p)	precision	p in **Fw.p** etc	p in **%w.pf** etc
dec	base 10	**I**	**i** and **d**
oct	octal	**O**	**o**
hex	hexadecimal	**Z**	**x** or **X**
setbase(b)	**b** is 8, 10 or 16		
endl	newline	**/**	**\n**
ends	append null		**\0**
flush	print buffer now		
setfill(c)	set fill character to **c**		
ws	skip whitespace on input		
setiosflags(ios::left)	left adjust		**%−**
setiosflags(ios::right)	right adjust		
setiosflags(ios::scientific)	exponential form	**E EN ES**	**e E**
setiosflags(ios::showpos)	show + sign	**SP**	**%+**
setiosflags(ios::showpoint)	print trailing zeros		
setiosflags(ios::fixed)	fixed point format		

Problem 8.8   *Consider the outputs from the following two programs. Try them, and see the effect of omitting the two **ios** flag settings.*

```
#include <iostream.h>
#include <iomanip.h>
#include <math.h>
main()
{
    for (int j=1; j<=200; j++)
    cout << j << " " << sqrt(float(j))
        << endl;
}
```

```
#include <iostream.h>
#include <iomanip.h>
#include <math.h>
main()
{
    cout << setiosflags(ios::showpoint);
    cout << setiosflags(ios::fixed);
    for (int j=1; j<200; j++)
    cout << setw(4) << j << setw(10) <<
        setprecision(4) << sqrt(float(j))
        << endl;
}
```

⊢ *8.4.1.1 Writing your own manipulators*

Facing **cout << setw(9) << setprecision(3) << x;** rather than **printf("%9.3f", x);**
you may agree that the supporters of **C++/C** standard I/O have a case. The way round
such inconvenience may be to declare your own manipulator. A manipulator is just a
function which returns a reference to **ostream** and whose first argument is also such
a reference. Suppose you use this **9.3** format a lot. You could define **myform** with

> **ostream& myform (ostream& s)**
> **{ return s << setw(9) << setprecision(3); }**

and then use it like this,

> **cout << " Answer " << myform << x << myform << y << endl;**

⊢ *8.4.1.2 The* **flush** *manipulator*

Your output is (probably) buffered: when you print a character it does not go to the
output at once, but the computer waits for a complete line and then sends them all at
once – which is more efficient and saves time. This is usually what you want. There
are exceptions, such as when you want to prompt for an input:

> **cout << " Please give destination: ";**
> **cin >> destination;**

This doesn't work properly as the message from the first line is buffered and
doesn't appear on the screen at the point when the input has to be made. Not very
user-friendly! An **endl** will clear the buffer, but that would move to the next line and
it wouldn't look so good. Instead you can use **flush** to clear the buffer, sending the
message to the screen.

> **cout << " Please give destination: " << flush;**
> **cin >> destination;**

8.4.1.3 **endl** *and the alternatives*

When you want to generate a new line at the end of a **cout** there are 3 ways of
doing this: by appending the character '**\n**', the string "**\n**", or the constant **endl**. The
first is not recommended as you may one day decide to add some words.

// Before adding helpful text		// After you add helpful text	
cout << x << '\n'	// OK	cout << x << ' is the value\n'	// wrong
cout << x << "\n"	// OK	cout << x << " is the value\n"	// OK

endl is actually superior to "**\n**" in that it is guaranteed to flush the output buffer.
Otherwise there is nothing to choose between them (they are both 4 keystrokes.) We
regard this as a matter of taste: our preference is to use **\n** if the last item is a text
string anyway so it's easy to include, and **endl** if it isn't.

★ 8.4.2   Formatting input with stream I/O

Formatting on input is much less important than on output. Nevertheless the **setw** manipulator is available and could be used, for example if you had a set of numbers that had been stored without intervening blanks to save space.

The **ws** manipulator is relevant for strings. Normally it is set, and initial *white space* – i.e. blanks, tabs, new lines – are ignored when reading a string. If you want to include them then this flag enables you to do so. Given an input file that starts with spaces (in this case 4 of them)

> **Fred**
> **8909.33**
> ...

Then
```
char s[50]
cin >> s;
```
will read the 4-character (plus null) string, **"Fred"**. To get the full 8-character string – if that is what you want – you add the line

```
cin >> resetiosflags(ios::ws);
```

and you would do well to set the flag again when you've finished.

8.4.3   Single-character I/O in C++

In addition to the << and >> operators there is another set of functions which can be used. (They can actually be used with any streams, we'll restrict ourselves to **cin** and **cout** for the moment but they're also very useful for file I/O.)

s.get(k) reads one character from stream **s** and stores it in **k**. (N.B. this is the same as reading one byte, apart from the carriage-return/line feed business). So if you press the 1 key in response to a **cin.get(i);** statement, then **i** will be given the value 49. There is a matching **put** for output: **cout.put(49)** will print **1** on the screen.

There is also a witty little function **s.putback(k)** which writes a character to the input stream, such that the next read attempt picks it up. It's useful if you're skimming characters till you find the one you want to start at: having found it you put it back, and then start serious input.

More than one character can be read in using **s.read(iarray , n);**, which reads *n* bytes into the array, and written with **s.write(iarray , n);**. Null characters are not treated specially with these two functions.

A variant of this which is more useful for inputting strings is **s.get(str , n , c);** where **str** is a string (array of **char**), **c** is the terminator character (which can be omitted and defaults to \n), and the string is then filled from the stream until $n - 1$ characters are read, unless the terminator character is met in which case it stops (leaving the terminator on the input stream); the null character is then added to the end of the character string.

8.5   A COMPARISON OF THE THREE METHODS OF I/O

To illustrate the differences in the systems consider a program to list prime numbers aligned in a number of rows of 10. The results we are aiming at are as follows:

2	3	5	7	11	13	17	19	23	29	10
31	37	41	43	47	53	59	61	67	71	20
73	79	83	89	97	101	103	107	109	113	30

Here is the version using **C++/C** standard I/O.

```
// prime listing using C++/C standard I/O
#include <stdio.h>
#include <stdlib.h>
int isprime(int r);       // function prototype
main()
{
    int ktest = 1 , nrows;
    printf(" Please give number of rows to be printed : ");
    fflush(stdout);                    // may not be needed on some systems
    scanf("%i" , &nrows);              // desired number of rows
    int  ntodo = nrows*10;             // at 10 values on each row
    printf("================================\n");
    for (int i=1;  i<=ntodo;  i++)
      {                      // loop over rows
         while(!isprime(++ktest)) ;             // find next prime
         printf("%5i" , ktest);
         if(i%10 == 0) printf("%8i\n" , i);     // end-of-line printing
      }
    printf("================================\n");
}
int isprime(int r)
   {         // is r a prime number? — used only for sequence from 1
      int static nprimes = 0;
      int static primes[201];                   // list of known primes
      for(int i=1;  i<=nprimes;  i++) if(  !(r%primes[i])  ) return 0;
      primes[++nprimes] = r;                    //  add to list of primes
      if (nprimes > 200) exit(EXIT_FAILURE);
      return 1;
   }
```

As a prime finder this code is inefficient (the programs illustrate formatting rather than mathematics). After the first prime, 2, we need only test odd numbers, not all numbers. Likewise we need to test prime factors less than the square root of the number. But the code as written is simpler – the invariable tradeoff in computing!

The C++ stream I/O equivalent program uses the same **isprime** function, which returns a 1 (true) if the number is prime, and a 0 (false) if it is divisible by one of its list of already known prime numbers. (The number 1 itself is not a prime number, according to the standard mathematical definition.) Note that it is not a general-purpose routine to test for prime numbers: it assumes that it is called for numbers in increasing order, not missing any out.

The request to specify the number of rows for printing may need to flush the output buffer to get the question output before the answer is expected. The version with standard I/O does this with the function call **fflush(stdout)**, while with stream I/O you output the **flush** manipulator.

The stream I/O version needs **iostream.h** and **iomanip.h** in place of **stdio.h**. Both need **stdlib.h** to get the **exit** function and the **EXIT_FAILURE**.

The banner line of equals signs (use more if you like) is used to show the start and stop of the output. This is easily coded using the **setfill** manipulator (which has to be reset afterwards). In **C++/C** standard I/O you write them all out, or do a little loop.

Another neat programming point (not connected with I/O) is the **while** loop with a null body – the action desired is all in the autoincrementing **++ktest**.

The main program is as follows:

```
// prime  listing using C++ Stream I/O
#include <iostream.h>
#include <stdlib.h>
#include <iomanip.h>
int isprime(int r);          // function prototype
main()
{
    cout << setw(40) << setfill('=') << "=" << setfill(' ') << endl;
    int ktest = 1 , nrows;
    cout << "How many rows to print? " << flush;
    cin >> nrows;
    int ntodo = nrows*10;
    for (int i=1; i<=ntodo; i++)
       {
           while( !isprime(++ktest) ) ;
           cout << setw(5) << ktest;
           if(i%10 == 0) cout << setw(8) << i << endl; // end—of—line printing
       }
    cout << setw(40) << setfill('=') << "=" << setfill(' ') << endl;
}

int isprime(int r)
   {   /* ... as before */    }
```

Here is the **Fortran** version. The problem here is the automatic new line after every print. This makes it harder to print the prime numbers one at a time as we go through. Instead we adopt a different philosophy, storing the primes in an array and printing them at the end, under the control of the **FORMAT** statement.

Note the compact **FORMAT** statement for printing the before and after banner of equals signs. A numbered **FORMAT** statement is used, to avoid having to specify it twice.

The array itself is printed with a double loop, and uses the way that the format is restarted when exhausted.

```
            PROGRAM primes_fort
            INTEGER :: i , j , ktest = 0
            INTEGER :: nprimes = 0
            INTEGER :: primes(1:200)
            INTEGER :: nrows , ntodo
            LOGICAL :: prime
            PRINT * , ' How many rows of primes (1 − 20)?'
            READ * , nrows
            ntodo = 10*nrows
            IF(ntodo > 200) STOP ' too many numbers for array size'
outer:      DO i=1 , ntodo
                prime = .FALSE.
middle:         DO WHILE( .NOT. prime)
                    ktest = ktest+1
inner:              DO j = 2 , nprimes
                        IF(MOD(ktest , primes(j)) == 0)  cycle middle
                    ENDDO inner
                    prime = .TRUE.
                ENDDO middle
                nprimes = nprimes+1
                primes(nprimes) = ktest
            ENDDO outer
            PRINT 100
100         FORMAT(1X , 40('='))
            PRINT '(10I5 , I8)' , ((primes(10*(i−1)+j) , j=1,10) , 10*i , i=1,nrows)
            PRINT 100
            STOP ' formatted primes all done'
            END PROGRAM primes_fort
```

The program logic all occurs within one unit, rather than having a separate function to test for prime numbers – it seems more natural this way. The ability of **Fortran** to jump out of nested loops (something **C++** can only achieve by using **goto**s) made this easier.

8.6   STREAMS AND RECORDS

The *stream-based* system of I/O used in **C++** and **C** (yes, even 'standard' **C++/C** I/O is stream-based) is fundamentally different from the *record-based* philosophy of **Fortran** (in all its various incarnations).

To understand the difference requires another journey into the ancient history of computers. Before the development of cheap and reliable magnetic media, programs and data were stored on paper tape or punched cards. The tape or card was passed between a light and a photocell, and holes punched in the tape or card represented binary digits to the computer.

Now the length of a paper tape is, like a piece of string, what you would like it to be, so an entire program or dataset would usually be stored on one roll of paper tape and the computer would read it all in one character at a time. The end of each line was specified by the NL (or CR–LF) character(s), and a line could be long or short, as required.

Each card, on the other hand, contained room for a fixed number of characters – 80, in fact. A program or dataset would be stored on a number of cards, and the computer would read in one card after another. Each line occupied a single card (even if it was only a short statement like **END**), which meant that there was no need for any line termination character to be used, and that each line was exactly 80 characters long.

These systems are long gone, but their ghosts remain with us. The system of stream I/O works one byte at a time, with special significance being given to the NL character. Record I/O assumes that characters are grouped into records of a certain length (80 or whatever) and operations are performed on a complete record.

The explains why, as you'll have noticed, you have to give the newline instructions explicitly in the **C++/C** examples we have used, which are stream based, but not in the record-based **Fortran** ones. **cout** << **"Hello"**; in **C++/C** outputs the 5 characters, as requested. **PRINT** * , **'Hello'** in **Fortran** outputs a whole record: the 5 bytes and the NL record terminator. (Or, if appropriate, 5 bytes and 75 blanks).

Suppose your program wants to input a value from a file containing

```
1 2 3 4 5
6 7 8 9 10
```

then a stream-based **READ** statement will read the first byte as 49, ASCII code for 1, then the second byte as 32, ASCII space, realise that it's got to the end of the number and return the desired value of 1. A record-based **READ** statement will read the entire line (=card), and then decode the first value as 1. So far no difference – the point is that for your *next* input the stream-based system will carry on at the next byte, returning with the value 2, but a record-based system will carry on at the next record, returning the value 6.

Fortran I/O is record-based. There is a rudimentary stream I/O offered, but it's still within the global record-based philosophy – indeed it has the alternative title of 'partial record I/O'.

8.7 FILES: STREAMS AND UNITS

A simple interactive program reads its input from your keyboard and writes its output to your monitor screen. Accordingly, all three styles include a *standard input* and a *standard output*, which are normally assigned to the keyboard and the screen. But you can override these assignments using the facilities of the operating system. In UNIX, for example, if the program **myprog** expects input from the keyboard and outputs answers on the screen, then the statement

myprog   < instructions   > results

will take input from the file **instructions** – perhaps prepared beforehand using your favourite editor, and write output to the file **results** – which you can browse through at your leisure.

In DOS the format is identical (though you're restricted to eight-character filenames and the three-character extension):

myprog   < instruct.in   > results.out

But a general program should be able to access many files, and other devices. We need to specify within a program whether a message is to be written to, say the screen, or the printer, or a file on disc. All three systems do this in the same basic way:

First the hardware destination is specified by means of an *open* command, which makes a connection between the physical object and a software entity (called a *unit* in **Fortran**, a *file pointer* in C++/C standard I/O, and a *stream* in C++ stream I/O).

Secondly, I/O operations (read, write and some others) are performed by the program using this connection.

Thirdly and finally, the connection is *closed*.

Fortran uses *unit numbers* to link the hardware and software entities. A unit number is an integer in the range 1 to 99. The **OPEN** statement establishes a unit number for the file or device. This is, within the program and while in use, unique. Subsequent **READ**, **WRITE**, and **CLOSE** statements refer to that unit number.

So to read data from a file – say it's called **measurements.dat** – you first choose a unit number (say 20) and issue an **OPEN** command; thereafter you can read data from unit 20. The **INQUIRE** command can report on all aspects of a file or a unit.

```
OPEN (20 , FILE='measurements.dat')
INQUIRE (20 , EXIST=ex , NAME=text)    ! etc What is on unit 20?
READ (20 , *) X , Y , Z
CLOSE (20)
```

OPEN, CLOSE, and **INQUIRE** are very powerful and comprehensive functions.

Because **Fortran** I/O is record based, the length of the records must be specified for each file. If you don't want the system default (for example, if you're printing a file and you know that all the lines are less than 20 characters long so that 80 character records would be a gross waste of space) you specify it with the **RECL** keyword. The units are system dependent so this needs care.

```
OPEN (UNIT=20 , FILE='measurements.dat' , RECL=20)
```

In the C++/C standard I/O, the software link is a *file pointer*. A type called **FILE** (yes, in upper case!) is defined in **stdio.h**; it's a structure containing the hardware and access details. In the program you actually use pointers-to-FILE. These are declared, and then the details are filled in by the function **fopen** which creates the **FILE** and assigns the pointer. So your program needs to contain the file definition and the actual opening before you can get started. In C++ stream I/O you also declare and define the connection – here referred to as a filestream – in two separate stages. **open** is a member function of the stream object: opening (and closing) are things that can be done with streams. You have to include the header file **fstream.h** if you haven't already done so.

```
FILE *pf;   // pointer to FILE
pf = fopen("measure.dat" , "r");
fscanf(pf , "%f %f %f\n", &x, &y, &z);
fclose(pf);
```

```
ifstream measurements;
measurements.open("measure.dat");
measurements >> x >> y >> z;
measurements.close( );
```

The actual input or output is done by **READ** and **WRITE** in **Fortran**, as discussed in §8.2.4.1, by the usual get-from and put-to chevron operators >> and << in C++ stream I/O, and by the functions **fscanf** and **fprintf** in C++/C stream I/O, which are like **scanf** and **printf** with the file pointer as the first argument. For single characters the **get** and **put** member functions of C++ stream I/O are used, and the C++/C **getchar** and **putchar** routines have equivalents **fgetc(FILE*)** and **fputc(int, FILE*)**.

Problem 8.9   *Write a program that will read numbers from the keyboard and write them to a file called* **data.store**.

The use of unique integers to denote file connections is a strength and a weakness, and both arise from the same cause: communication between routines (and between programmers). If, in an enormous **Fortran** program written by a team of programmers, it is decided that subroutine **init** – written by Janet – will open a file, and that subroutine **process** - written by John – will read data from it, then Janet merely has to tell John the unit number (say, 22) that she's chosen to use for her **OPEN** statement, and he then uses that for his **READ** statements. In C++ – with either method – she has to ensure that her file pointer or stream is accessible to others.

Problem 8.10   *How could this be done (a) using global variables, (b) by defining the file as an object, and (c) by returning the file pointer as the function return value?*

The problem is that nobody else working on this large program can now use unit 22 for their purposes. That's no limitation as there are still 98 other possibilities – but the project management has to make sure that the rest of the team know.

Standard input and standard output in **Fortran** are merely two pre-assigned units. The exact numbers depend on your system. Sometimes they are 1 and 2, very often they are 5 and 6.

8.8   OPENING A CONNECTION

The **Fortran OPEN** statement is **OPEN (unit, parameter list)** where the optional parameter list is a set of **name = value** pairs, given in any order. Many have the form *keyword=characterstring* delimited by quotes. There are many options in **Fortran** such as **ACTION=, ACCESS=(DIRECT , SEQUENTIAL)**, (file) **STATUS=(OLD , NEW , UNKNOWN , REPLACE , SCRATCH) , ERR , IOSTAT**, and more. This wide-ranging file-handling flexibility is a strength of **Fortran**. For example,

```
OPEN(22 , FILE='data.01' , ACTION='READ')      ! Open existing file for input
OPEN(23 , FILE='results.01' , ACTION='WRITE')  ! Open a file for output as unit 23
OPEN(FILE='log' , ACTION='WRITE' , POSITION='APPEND' , UNIT=24)
MYUNIT = 25
DFILE = 'dictionary.dat'
OPEN(MYUNIT , FILE=DFILE , ACTION='READWRITE' , IOSTAT=ios)
```

If a unit is used without being opened then the system will open a default; the filename depends on the system but is often (for, say, unit 22) something like **FOR022.DAT** or **fortran.022**. This is not a feature any decent programmer would want to take advantage of.

In **C++/C** standard I/O the **fopen** function returns the file pointer as the result. It has only two arguments. The first is the name of the file (as a character string – this can be a constant or a variable). The second is also a character string which specifies the various details of how you want to handle the file. Each option is a single character (but remember that even if there's only one character it's still a string and needs double quotes, not single quotes!). For example,

```
FILE *fp1 = fopen("data.01" , "r");      // open existing file for read
FILE *fp2 = fopen("results.01" , "w");   // open new file for write
FILE *fp3 = fopen("log" , "a");          // open file for output,
                                         // appending data at the end
```

With **C++** stream I/O, you first define a filestream. There are 3 types: **ofstream** for output, **ifstream** for input, and **fstream** for files that you want to use for both. A filestream object has member functions **open** and **close** defined for it. Because input and output streams are defined separately, you don't need to specify this by an argument in the open statement. Other details – if necessary – can be specified by the second (optional) argument, an integer which can be built out of bits defined in **fstream.h**. (If you want to specify more than one, they must be ORed together.)

```
#include <fstream.h>
ifstream f1; ofstream f2; fstream f3;
f1.open("data.001");                        // input file stream
f2.open("results.001");                     // output file stream
f3.open("dictionary.dat" ; ios::in | ios::out);
```

The various options for the open statements for the three systems are discussed in the appropriate forthcoming section.

8.8.1   Error trapping

It's surprising how often your attempts to open a file are unsuccessful. Perhaps because you got the name wrong (spelling, upper- and lower- case differences, simple misunderstandings), or you forgot which directory you were in, or the operating system bureaucracy doesn't give you the appropriate access permission to do what you want. So it's vital that you test any file opening for success.

The **Fortran OPEN** statement provides two ways of doing this: the **IOSTAT** variable and the **ERR** label. (Notice that these are slightly different from the usual *keyword=characterstring* format.) Either or both of these can be specified (and at least one of them should be!). The **IOSTAT** variable is set to zero if the file was opened successfully; if it failed then it's set to a positive value. (The meaning is system dependent.) If the **ERR** label is given, then on return from an unsuccessful **OPEN** statement the program will jump straight to the label given. As discussed in §8.2.4.1 this is at best the lazy programmer's alternative.

```
      OPEN(21 , FILE="old.data" , ACTION='READ' , IOSTAT=J)
      IF(J /= 0) THEN
         PRINT * , ' Failed to open OLD.DATA file '
         PRINT * , ' Error number ' , J , ' Consult Manual!'
         STOP
      ENDIF
      OPEN(22 , FILE="new.data" , ACTION='WRITE' , ERR=999 , IOSTAT=J)
      RETURN
999   PRINT * , ' File opening error  ', J
      STOP
```

In **C++/C** standard I/O, an unsuccessful call to **fopen** returns a NULL file pointer, which can and should be tested for. (**exit()** requires the header **stdlib.h**.)

```
      FILE *fp1 = fopen("OLD.DATA", "r");
      if( fp1 == NULL )     // or if( !fp1 )
      {
         printf(" Failed to open OLD.DATA file \n");
         exit(EXIT_FAILURE);
      }
```

In C++ stream I/O, an unsuccessful **open** means that the stream is not established. It is then – when treated as a pointer – zero, which can be tested for.

```
      ifstream datafile;
      datafile.open("OLD.DATA");
      if(datafile)     { /* ... */ }
      else
      {
         cout  <<  " Fail to open OLD.DATA\n"; exit(EXIT_FAILURE);
      }
```

★ 8.8.2   Reading or writing with old or new files

Some files you open because you're going to want to read data from them. Others you open because you're going to write data to them. Some for both reasons. It's important to give this information when you open the file, because of hardware details such as caching (using temporary fast-access storage).

When you open a file because you want to read data from it, then that file must exist already, and if it doesn't then an error should result. If you open a file because you want to write data to it then it's not so clear cut: usually you want to write to a new file, but sometimes you'll want to write to an old one, perhaps by appending data at the end. And even with a new file, what should happen if the system discovers it exists already? It may be that you're happy to overwrite the old file with a new (better) one, or it may be that these are useful results that you don't want to overwrite and lose.

The three systems cope with this in similar ways. In **Fortran** there is a keyword called **STATUS**: this can be set to **OLD** if you insist that the file already exists, **NEW** if you insist that it doesn't. If you want to create a new file, and overwrite any previous version (if there is one) then it should be set to **REPLACE**. There is no way to say (with a single **OPEN** statement) that you want to use an existing file if there is one and create a new one if there isn't. The default is **UNKNOWN**, but what that does is system-dependent and therefore dangerous. There is another keyword called **ACTION** which is set to **READ** if the file is only to be used for reading, **WRITE** for writing, and **READWRITE** in cases where you will read and write to the file.

In **C++/C** standard I/O the file status requirement is driven by the read/write specifier. If the option character string in the **fopen** statement is set to **"r"** for read, then the file must exist. If it is set **"w"** for write then a new file will be created, overwriting any earlier version. With **"a"** for append then a new file will be created if necessary, but if one already exists then that one will be used. The fun comes when you want to open a file for both reading and writing – as sometimes happens, for example in a database program. (See the later section on random access.) This is specified by adding a + to the character string. Then **"r+"** and **"w+"** both mean that a file is to be opened for reading and writing; the difference is in the file status on opening. **"r+"** opens a file which must exist, and **"w+"** opens a new file.

In **C++** stream I/O the status is controlled by the bit string in the open function arguments. There is a bit **ios::nocreate** which causes the opening to fail if the file doesn't exist (i.e. it will not create a file for you) and **ios::noreplace** which causes failure if it does (i.e. it will not replace an existing file). The read/write status follows from whether the stream is an **istream** for input or an **ostream** for output. The bits **ios::in** and **ios::out** are available for opening files which will be read from and written to. These should be defined as **fstream** objects rather than **ifstream** or **ofstream**.

Problem 8.11   *Translate into the two* C++ *systems the* **Fortran**
OPEN(20 , FILE='data.016' , STATUS='OLD' , ACTION='READ')

★ 8.8.3   Formatted (text) and unformatted (binary) data

Writing data to a file which will be read by another program is different from writing it to a file (or screen) which is going to be read by a human. For a 4-byte integer (say) you just want to transfer the 4 bytes from memory onto 4 bytes on the disc. You don't need to convert it into a stream of ASCII characters. It's simpler, it's faster, it's almost always more compact, and it avoids the danger of loss of precision. Such dumping of the binary data in memory straight to a file is called *binary* or *unformatted* (not to be confused with 'free format') I/O.

A particular file will be used for formatted or for unformatted I/O but not for both – so you specify your intentions when you open it. With **Fortran** this is done by **FORM='UNFORMATTED'** or **FORM='FORMATTED'** as desired in the **OPEN** statement. (The default is **'FORMATTED'**, except for random access files – §8.9.) For **C++/C** you put a **b** for binary in the option string of the **fopen**. For **C++** stream I/O you set the bit **ios::binary** in the optional word.

```
OPEN(12 , FILE='raw.data' , FORM=UNFORMATTED , RECL=50)
fp2 = fopen("raw.data" , "b");
f.open("raw.data" , ios::binary);
```

For **C++** this is just a way of treating a file, not a property of the file itself. You can write a formatted (text) file and read it back as unformatted (binary). **Fortran** files may contain information that specifies whether they're formatted or not.

Having opened the file you need to read from or write to it. In **Fortran** this is easy. You just omit the format specification. If units 10 and 11 have been opened for unformatted (binary) read and write, then you just say

```
READ(10) X , Y , Z          |          WRITE(11) X , Y , Z
```

to transfer the 3 floating-point values from and to the files. Note that it writes a record, and the record length of this file must be big enough to hold 3 numbers.

The equivalents in **C++/C** standard I/O are the routines **fwrite** and **fread**. These have 4 arguments **fwrite(address , size , number , file—p)**; **file—p** is file pointer, opened as usual. **number** is the number of items that will be written, and **size** is the size of each item (so it's often specified using **sizeof**). **address** is the address of the word in memory from which transfer will occur – it can either be a variable preceded by the address-of operator **&**, or the name of an array.

In **C++** stream I/O member functions **read** and **write** do very much the same thing. Their arguments are the address of the data and the number of bytes. The address must be a pointer-to-char – but if you want something else then a cast will fix it.

```
int i;                              #include <fstream.h>
float a[10];                        ofstream f;
fp = fopen("results" , "wb");       f.open("results" , ios::binary);
fwrite(&i , sizeof(int) , 1 , fp);  float data = 12354.6;
fwrite(a , sizeof(float) , 10 , fp); f.write( (char*) (&data) , sizeof(data) );
```

★ 8.9   MOVING ABOUT

Most files, like novels, are read by *sequential access*, starting at the beginning and proceeding through to the end, and written in the same way. But some files are like dictionaries, and you want to skip forwards and backwards as you read (or write) them; database files are a typical example. This is called *random access* or *direct access*, as opposed to the more usual *sequential access*. Actually the term 'random access' is a bit of a misnomer – you access where you want, not 'at random' at all. ('Direct access' isn't much better as it implies some accesses are indirect.)

★ 8.9.1   Moving about with sequential files

Although a sequential file is normally treated by starting at the beginning and proceeding to the end, there are a few rudimentary commands that move about in it.

One is to go back to the beginning of the file: if you have read through some data processing it you may well want to go back and do it again with some improved parameters. In C++/C and **Fortran** this is called a *rewind* – because these languages still think in terms of files on magnetic tape!

rewind(fp); | REWIND(22)

C++ stream I/O does not contain this explicitly: you use the **seekp** or **seekg** member functions (described in the next section) to set the position to zero.

You can also move to the end of the file with **ENDFILE(unit)**. This can be useful if you are writing to a log file; you want to deposit messages from your program but you don't want to overwrite stuff that's already been put there. Another way to do this is by specifying the position in the **OPEN** statement.

OPEN(30 , FILE='jobs.log' , ACTION='WRITE' , POSITION='APPEND')

or in C++/C the **a** option is like **w** for write, except that data is added at the end:

FILE *fp = fopen("jobs.log" , "a");

For **C++** stream I/O there are two bits that can be set in the options word in the **open** statement, **ios::app** ('append') means that data will be written to the end of the existing file, and **ios::ate** ('at end') means that the pointer is initially set to the end of the file.

Remember that if you want to specify two or more of these bits, you just **OR** them together

results.open("run97.data" , ios::binary | ios::ate);

Fortran also provides **BACKSPACE(unit)** which backs up one record, so that the last **READ** statement can be done again, presumably with a different format. The C++ stream I/O **putback** (see §8.4.3) is similar though it works with items rather than records.

★ 8.9.2   Random access

For **Fortran** this use must be specified when the file is opened by the keyword-value pair **ACCESS='DIRECT'**.

OPEN(67 , FILE='database.dat' , ACCESS='DIRECT')

In the two **C++** systems random access fits in more naturally. But you should specify the file as binary, otherwise there will be confusion between counting characters and counting bytes (that old newline problem again!) on some systems.

Moving about in the file, the **Fortran** system is based on records whereas the **C++** systems count in bytes. (Of course.) In **Fortran** you use the usual **READ** and **WRITE** statements and you also specify the record number:

! Formatted read	! Unformatted write of 10 values
READ(67 , '(I8)' , REC=94) J	DO j=1,10
	WRITE(68 , REC=j+17) value(j)
	ENDDO

Free format (e.g **READ(67 , * , REC=94) J**) is not allowed, unfortunately.

You can access a file by both direct access and sequential access – but not both at the same time. You have to deal with it in one way, then close the file, and reopen it for the other type of access.

C++/C standard I/O moves around using a function **fseek(file–pointer , offset , origin)** which sets the file position to the desired value. The offset is counted in bytes, (it's a long integer, by the way) so you have to do more arithmetic than in **Fortran**! The **origin** specifies where the offset is counted from. There are 3 possibilities: **SEEK_SET** is the usual one, and counts from the start of the file. **SEEK_CUR** counts from the current position. **SEEK_END** counts from the end (so the offset must be negative). The value returned is zero unless there is a problem, so it can be used to test that the operation worked successfully. There is a complementary function **ftell(file–pointer)** which returns (as a long integer) the value of the current position.

C++ stream I/O keeps account of two positions for a file, the *get position* for reading and the *put position* for writing. There are thus two member functions for changing the positions, **s.seekg(k)** sets the get position to **k** for random-access read, and **s.seekp(k)** is used for random-access write. The normal origin is the start of the file, if you want to use the current point or the end, you use the optional second argument, setting it to **ios::cur** or **ios::end**.

There are also two functions to tell you where you've got to, **s.tellg()** which returns (as a type **long** integer) the get position for stream **s**, and **s.tellp()** which returns the put position.

For example, suppose a library catalogue has details (number of pages, title, and author, for instance) for many books stored on a direct-access file. Suppose the data is written as a 4-digit integer, a 20-character string and a 100-character string respectively. Then to get the 100th entry (say) you would use statements like

```
#include <stdio.h>
int pages;
char name[20] , title[100];
int recsize = sizeof(pages)
    + sizeof(name) + sizeof(title);
FILE *fp=fopen("library" , "rb");
fseek(fp , 99*recsize , SEEK_SET);
fscanf(fp , "%4i , %s20 , %s100)",
    &pages , name , title);
```

```
INTEGER :: PAGES
CHARACTER :: NAME*30, TITLE*100
OPEN(10 , FILE='cat', &
    & ACCESS='DIRECT' , RECL=124)

READ(10 , '(I4 , A20 , A100)' ,&
    &REC=100)   pages , name , title
```

Problem 8.12   *Rewrite the* **C++/C** *standard I/O example above using* **C++** *stream I/O.*

If you open a new file for direct access you can then write records (or bytes) of any number you choose (within hard limits such as the size of the disc.) You could open a file, write record 1 and record 1000, and then stop. When you open an existing file you can write records that are on the file – here all the records 1 to 1000 will have been allocated even though only 2 have been written, so they can all be used – but you are liable to hit a problem if you try and go outside the existing range of records.

8.10   CLOSING THE CONNECTION

The close statement is an important one. It flushes any internal buffers to the file, and lets the operating system know that the file is available for other users. (Operating systems are usually reluctant to let you look at a file which another program is processing, and quite right too!) This closing is usually done automatically at the end of a job, but it's good practice to make sure by doing it yourself.

The **Fortran CLOSE()** command must give the unit number and may also specify the arguments **IOSTAT**, **ERR**, and **STATUS=(KEEP, DELETE)**.

```
CLOSE(20 , IOSTAT=close_status , STATUS='KEEP')  ! in Fortran
fclose(fp1);               // in C++ standard I/O
results.close( );          // for file output stream 'results' in C++ stream I/O
```

Closing the connection also frees the file identifier (**20, fp1** or **results**, as appropriate) for use within your program – you may perhaps want to process many files of data using the same unit number, or file pointer, or file-stream object.

8.11 GRAPHICS

Although graphics is of great and increasing importance, it is not part of the specification of either language. If you want to write programs that draw pictures you will have to obtain an additional graphics library that matches your equipment. Although most graphics is done using bit-mapped devices – screen and laser printers – many graphics packages continue to be largely based on pen-and-paper ideas. A typical package will contain items like change-pen-colour, move-to-point, draw-to-point, all of which assume that there is a pen being moved around the paper, sometimes in contact with it, sometimes not. Your package may have an explicit draw-circle or draw-polygon command in it, or it may not.

Most of the headaches with graphics packages come from coordinate transformations. Your output device will have some native coordinate units, origin, orientation, and window. For example, if you have an HP7470 plotter with its DIP switches set to A4 mode, then it uses plotter units of 0.025 mm, the origin is at the top left-hand corner of the page, the x-direction is vertically downwards and the y-direction is to the right, and the plotting area goes from 0 to 10900 units in x and 0 to 7650 units in y. If you have a postscript device then the basic unit is 1/72 of an inch, the origin is at the bottom left, the x-direction points right and the y-direction points up. If you have some other device then the numbers and directions will be different but the ideas will be the same.

You want, say, to plot the current (from 0 to 20 milliamps) in a specimen with voltages in the range 0 to 50 volts across it. And you'd like your origin in the bottom left-hand corner (but not quite at the edge, to leave room for titles and axis units) with volts going right and current going up.

Now there are two extreme ways of going about this. You can do all the transformations yourself and work entirely in the plotter system. Or you can specify the scales, and then do your plotting in a civilised and convenient system. It's a lot easier to do the latter – and in many cases your graphics system will provide library functions to do this.

The most general transformation between your units and the hardware units will include two scaling factors, a rotation, and two offsets. It's important to get the order of these right! A rotation followed by a shift is not the same as the shift followed by the rotation! Very often the first attempt at a graphics picture is completely empty. Two common reasons for this are either (a) you have rotated and shifted the coordinate system such that your image is outside the visible region or (b) your have set the scale factor wrong, and your image is microscopically small.

Commercial software exists for various platforms and it is well worth while searching the Web. An excellent, free, comprehensive, implementation is called **PGPLOT** (conceived and created by Sze Tan a decade ago) which is maintained at **http://astro.caltech.edu/~tjp/pgplot/ver511.html**. The graphics functions can be called directly from **Fortran 77** and **Fortran 90** programs. A C binding library allows **PGPLOT** to be used with **C++/C** programs.

9

Numerical Methods

After an initial and disquieting example, the details of the IEEE 754 system for storing floating-point numbers are presented, for both single and double precision. That enables us to explain sources of numerical error that arise in floating-point calculations, and how to avoid them. As examples we fit straight lines and then higher polynomials.

In this chapter we look at some aspects of numerical computation and highlight some pitfalls and problems. The problems occur because when computers do calculations they work with *finite precision*; numbers are not stored exactly, only to some number of significant figures, perhaps just 6 or 7 places. Even if this seems ample for your input values and final results (and few of us really need to quote answers that accurately), it may happen that you need more precision than you realise in your intermediate calculations. When this occurs you can either rethink your algorithm, or tell the compiler that you need to work to higher precision. The pitfalls occur because this may happen without your realising it. Even although computers are powerful tools they respond mindlessly to the instructions we have given them; a major part of the science of programming is ensuring that what we program is what we actually intended.

9.1 EXPLORING FLOATING-POINT NUMBERS

The real number line, as used in mathematics, contains infinitely many rational numbers, even on the range $(0,1)$, and between any pair of rational numbers, no matter how close, lie infinitely many irrational numbers. In computer arithmetic we use floating-point numbers of finite precision as the best available substitute for the infinitely precise real numbers of mathematics. Much of the time there is no practical difference between the two. But sometimes it can give us a (nasty) surprise.

We start with some simple fractions: testing whether $1./3. + 2./3. = 1$, and whether 0.1 is the same as $\frac{1}{10}$. Then some algebra: is $(x + y) + z$ the same as $x + (y + z)$?

```
PROGRAM FP_numbers
! 1/3+2/3 is not 1 and 0.1 is not 1/10
REAL :: third = 1.0/3.0 , one = 1.0
REAL :: twothirds = 2./3.
CHARACTER :: fmt*11 = '(A , E12.8)'
PRINT fmt , ' 1./3.+2./3.−1. '&
        , third + twothirds − one
IF( 0.1 /= one/10.0 )   PRINT * , &
    ' 0.1 − 1.0/10.' , 0.1 − one/10.0
END PROGRAM FP_numbers
```

```
PROGRAM non_associate
! to show (x+y)+z /= x+(y+z)
REAL :: x=1e8 , y=−1e8 , z=1.0
REAL :: xy , yz
xy = x + y; yz = y + z  !store result
PRINT * , ' xy , yz , xy+z , x+yz ,'&
     & , ' supposedly equal, not 1. and 0.'
PRINT * , xy , yz , xy+z , x+yz
END  PROGRAM non_associate
```

Problem 9.1   *Try this on your system, in whichever language you prefer.*

We found that **third** + **twothirds** − **one** gave 2.98...E−8 rather than zero, and that **0.1−one/10.0** gave 1.49...E−9. Which shows again that you should

---

Never test floating-point numbers for equality!

From the second program we found that $(10^8 - 10^8) + 1.0 = 1.0$ but that $10^8 + (-10^8 + 1.0) = 0.0$, which is not the usual law of arithmetic.

There is nothing unusual about these results, except their unfamiliarity. This knowledge is an important part of our computational skills.

Typical systems use 32-bit *single precision* FP representations and/or 64-bit *double precision*. Or a mixture: the registers used in arithmetic calculations often use double precision, so even though single precision values are loaded from and stored in memory, our calculations may not proceed quite as planned. Indeed, if the calculation in the second program were done in-line, as **PRINT * , xy , yz , (x+y)+z , x+(y+z)** this does produce 1.0 and 1.0. This is because more accuracy is used in the intermediate arithmetic registers than is used in storage. (You can still show the effect, however, by using 1.0e20 instead of 1.0e8.)

9.1.1   Relative accuracy with floating-point numbers

Only some numbers are represented exactly in FP representation. In a typical system they range from $\pm 1.18 \ldots 10^{-38}$ to $\pm 3.40 \ldots 10^{+38}$, with the addition of 0.0 (zero). The spacing between successive representable numbers varies in a roughly logarithmic fashion and is $1.19 \ldots 10^{-7}$ at 1.0000000 and $\approx 10^{31}$ near the maximum. Most reals have a non-terminating binary representation. Even an apparently standard value such as 0.1_{10} is 0.0 0011 0011 $\ldots_2$ and it can only be stored approximately (just as $\frac{1}{3}, \frac{1}{6}, \frac{1}{7} \ldots$ cannot be written down exactly as decimals). If the number is encoded in 32 bits with 8 for the exponent and 24 for the *significand** then *rounding errors* of about 1 in 2^{24} would be expected, which is about 6 in 10^8, and indeed, the accuracy of FP numbers varies between 6 and 9 significant decimal digits. Accuracy here means that if you translate a decimal into a binary of fixed length, and back again, then the difference is the accuracy. A good estimate is 1 in 10^7.

The relative accuracy *epsilon* is the smallest real **eps** for which **1.0 + eps** is distinct from **1.0**. The following programs evaluate the smallest number **eps** (assumed to be a power of 2) such that **1.0 + eps /= 1.0** or **1.0 + eps != 1.0** as appropriate.

```
#include <iostream.h>
#include <math.h>
main()
{    //program epsilon
    float eps , one=1.0 , oneeps;
    eps = pow(2.0 , −14);    // as a start
    for (int k =1; k <= 16; k++)
    {
        eps /= 2.0; oneeps = one + eps;
    int testeps = (oneeps != one);
        cout << oneeps − one <<" "<< eps
            <<" "<< testeps << endl;
    }
}
```

```
PROGRAM epsilon
REAL       :: eps , one=1.0 , oneeps
INTEGER  :: k
LOGICAL  :: testeps
eps = 2.0**(−14)       ! as a start
DO k = 1 , 16
    eps       = eps / 2.0
    oneeps   = one + eps
    testeps  = (oneeps /= one)
    PRINT * , ' one+eps , eps ,' , &
    ' testeps ', one+eps , eps , testeps
ENDDO
STOP
END  PROGRAM epsilon
```

Problem 9.2   *Code this and try it. The output of the logical comparison* **testeps** *is T or F in* **Fortran** *and 1 (true) or 0 (false) in* **C++**.

The result we obtained from this program was that **eps** is $1.192093E-7$ or 2^{-23}.

Problem 9.3   *What happens if you program instead in one line (as it is very easy to be tempted to)* **epstest** $= (1.0 + \mathbf{eps}) - 1.0$? *Why?*

Problem 9.4   *Now try the original program, but testing* **one** − **eps** *instead of* **one** + **eps** *for equality with* **one**. *Explain the (surprising) result – you'll need the details of the FP system in* §9.2.

---

* The significand is also called the *mantissa* and the *fractional part*,§3.2, §9.2.

9.1.2 Theory is all very well – what about *my* machine?

Each language contains details of the FP model which has been implemented on the hardware in use, accessible by the programmer.

In **Fortran** various *inquiry functions* return the required information. Suppose you are using the number **f = 123.56e0**. This may or may not be represented exactly. Probably not.

The gap between representable numbers of this magnitude is **delta = SPACING(f)** which is $7.63 \cdot 10^{-6}$ on our system. (The relative spacing caused by a change in the smallest bit varies between 10^{-6} and 10^{-8} because of the floating-point representation.)

The system function **NEAREST** can be used to find the nearest neighbours of **123.56e0**, i.e. the two numbers definitely above or below **123.56e0** which can be represented exactly. (This actually means that they are *not* the nearest number, as the compiler cannot tell whether that is above or below your value.)

The closest larger number which can be exactly represented is

f_more = NEAREST(f , +1.0)   and our system gave 123.5600052.

The closest smaller number which can be exactly represented is

f_less = NEAREST(f , −1.0)   and our system gave 123.5599899.

Notice that the difference between these two is just twice the spacing, as it should be. The nearest representable number to **f** is hence **f_less + delta = f_more − delta**.

Other functions you may use are given in Table 9.1, and discussed in considerable detail in Kerrigan (1994, Chapter 9).

Table 9.1 The numeric inquiry functions in **Fortran**

Function	Purpose
These functions give information about numbers of the same **KIND** as **x** (real or integer)	
DIGITS(x)	number of significant binary digits
EPSILON(x)	value of *eps* : largest value that $1.0 + eps \ /= 1.0$
HUGE(x)	largest absolute value **x** can hold
MAXEXPONENT(x)	integer: the maximum exponent of **x**
MINEXPONENT(x)	integer: the minimum exponent of **x**
PRECISION(x)	number of decimal digits guaranteed
RADIX(x)	the base used (probably 2)
RANGE(x)	range guaranteed available, as a power of 10
TINY(x)	smallest absolute value **x** can hold
These functions give information about a particular **x** (which must be real)	
EXPONENT(x)	integer: exponent of **x**
FRACTION(x)	real: fractional part of **x**
NEAREST(x , s)	closest exact machine real to **x** in the direction of $s = \pm 1.0$
RRSPACING(x)	reciprocal of the relative spacing of numbers near **x**
SET_EXPONENT(x , i)	returns the significand of **x** with its exponent set to **i**
SCALE(x , i)	returns **x** with its exponent part changed by **i**
SPACING(x)	absolute spacing of numbers near **x**

In **C++ float.h** contains maximum and minimum values, etc. A typical version is shown here. **FLT** and **DBL** refer to single and double precision, as described in the next section. #**define** is a pre-compiler command. It means that, if you have the line #**include** <**float.h**> in your program, any use of (say) **FLT_DIG** is equivalent to **6**.

// Typical float.h	from the GNU C Compiler: used in C++/C	
#define FLT_DIG	6	// Number of decimal digits
#define FLT_EPSILON	1.19209290e—07F	
#define FLT_MIN_EXP	(—125)	
#define FLT_MIN	1.17549435e—38F	// Minimum float
#define FLT_MIN_10_EXP	(—37)	
#define FLT_MAX_EXP	128	
#define FLT_MAX	3.40282347e+38F	// Maximum float
#define FLT_MAX_10_EXP	38	
#define DBL_MANT_DIG	53	
#define DBL_DIG	15	
#define DBL_EPSILON	2.2204460492503131e—16	
#define DBL_MIN	2.2250738585072014e—308	// Minimum double
#define DBL_MIN_10_EXP	(—307)	
#define DBL_MAX_EXP	1024	
#define DBL_MAX	1.7976931348623157e+308	// Maximum double

The **math.h** header file contains all the available mathematical constants:

// Part of math.h	Typical constants defined in C++/C from the GNU C compiler		
#define M_E	2.7182818284590452354	//	e
#define M_LOG2E	1.4426950408889634074	//	log 2e
#define M_LOG10E	0.43429448190325182765	//	log 10e
#define M_LN2	0.69314718055994530942	//	log e2
#define M_LN10	2.30258509299404568402	//	log e10
#define M_PI	3.14159265358979323846	//	pi
#define M_PI_2	1.57079632679489661923	//	pi/2
#define M_1_PI	0.31830988618379067154	//	1/pi
#define M_PI_4	0.78539816339744830962	//	pi/4
#define M_2_PI	0.63661977236758134308	//	2/pi
#define M_2_SQRTPI	1.12837916709551257390	//	2/sqrt(pi)
#define M_SQRT2	1.41421356237309504880	//	sqrt(2)
#define M_SQRT1_2	0.70710678118654752440	//	1/sqrt(2)

so that, after including the header file, we can write

    const double long_pi = M_PI;   const float   piby2 = float(M_PI_2);

and know that our variables will have full accuracy.

9.2   THE IEEE 754 FLOATING-POINT SYSTEM

Modern compilers adopt this FP system if the hardware allows, and all the **C++** and **Fortran** compilers we used do so.

9.2.1   Single precision: 32-bit numbers

A standard *single-precision* FP number (a **float**) occupies 32 bits. The leftmost is a sign bit s, the next 8 store the exponent e, with a bias of 127, and the last 23 store the binary fraction f. The normalised significand has the value $1.f$ in binary, i.e. the most significant bit is always set (and since we know this we don't need to store it!). Thus the 23 bits of f, plus the implicit bit, store 24 bits of accuracy. The range of e is from 1 to 254 (0 and 255 are special cases, as will be discussed shortly) so 2^{e-127} has the range 2^{-126} to 2^{127}. Thus the FP number has the value

$$(-1)^s \times 2^{e-127} \times 1.f$$

The maximum number that can be represented has all the bits of f set to 1 so that $1.f$ is nearly 2 and $xmax = 2^{128}(1 - 2^{-24}) = 3.402823 \cdot 10^{+38}$.

The minimum normal number has $f = 0$ and so is $xmin = 2^{-126} = 1.175494 \cdot 10^{-38}$.

At this point *progressive underflow* begins. (We discuss this on intellectual grounds only: do not try this at home!) The system uses the numbers with $e = 0$ to give a graceful loss of accuracy for numbers which are *less than* the minimum (normal) number. The idea is sensible: represent one of these numbers as

$$(-1)^s \times 2^{-126} \times 0.f$$

with the leading bit now a zero and allow the fraction f to decrease steadily to increase the lower range at the cost of *steadily reducing precision*. There are theoretical and practical reasons for choosing this over a *store zero* policy (strict underflow) which relate to proofs of how certain FP calculations will behave. The smallest such number has a single 1 in the 23rd place of f and represents $2.0^{-126} \times 2.0^{-23} = 2.0^{-149} \simeq 1.40\ldots 10^{-45}$. It is totally inaccurate! The compiler will not allow you to compute such numbers directly, e.g. as **2.0∗∗(−149)**, but you can reach them by stealth. However, *it is most ill-advisable to work in this range*, since the precision is out of control, and we discourage the idea strongly.

The number zero has $e = 0$ and $f = 0$, and s can be $+1$ or -1, which means that zero has a sign! The compiler interprets ± 0 in the same way. The number **INF** has $e = 255$ and $f = 0$. **+INF** is used to mark *overflow* when an attempt is made to store a number whose magnitude is too great to be represented, and **−INF** corresponds to *underflow* when a magnitude is too small. When $e = 255$ and $f \neq 0$ (at least one bit of f is set to 1) then the result is Not-a-Number, often printed as NaN. There is a large number of these, all different.

9.2.2 Double precision: 64-bit numbers

The IEEE standard defines another format contained in 8 bytes or 64 bits, with a 52-bit fraction f, an 11-bit biased exponent e, and a sign bit s. **xmax** $\simeq 2^{+1024} \simeq 10^{+308}$ and **xmin** $= 2^{-1022} \simeq 10^{-308}$, while **eps** $= 2^{-52} = 2.22\ldots 10^{-16}$. The value of the number is $(-1)^s \times 2^{e-1023} \times 1.f$.

You can specify this precision in your programs by the appropriate declaration. In C++ **double** is a standard type, an alternative to **float** (and **long double** is available on some systems). In **Fortran DOUBLE PRECISION** is a (discouraged) alternative to **REAL**. In both languages constants are specified as follows in scientific notation:

float x2 = 6.666667.e−1F;	REAL :: b = 2.126e0
double y1 = 1.42857142857143e−1;	DOUBLE PRECISION a=1.23132411d0

However, the language definitions have to apply to all platforms, even those which don't use IEEE format. Or 32-bit words for that matter. The legal definition of **DOUBLE PRECISION** is that it provides a better precision than the standard, but it is not guaranteed how much better. **double** is only guaranteed to be no worse than the standard. If you need to write code that is portable and future-proof then this matters. The **Fortran** language provides a way of doing this, using **KIND** (see also §3.2.1). The **KIND** of a constant or variable is an integer which specifies the representation being used. For floating points, **KIND=1** probably represents the 32-bit IEEE representation and **KIND=2** the 64-bit. So you write equivalently

REAL (KIND=2) :: range , pressure	DOUBLE PRECISION :: range , pressure
CALL rootsub(1.2_2)	CALL rootsub(1.2D0)

but this is not really safe as there is no guarantee that this equivalence will hold in all systems, forever. For portable code you can use the functions **KIND** and **SELECTED_REAL_KIND**, which are part of the standard language. **KIND(x)** returns the **KIND** value of the argument. So a safer version of the above would be

```
INTEGER, PARAMETER :: dble=KIND(1.0d0)
REAL (KIND=dble) :: range , pressure
CALL rootsub(1.2_dble)
```

This gets away from that hardwired **KIND=2** though it still leaves you at the mercy of unexpected double-precision implementations. For a completely bullet-proof program you first have to decide on how many decimal digits and how big an exponent range you really need – let's say 12 digits and a range $10^{\pm 60}$ – and code:

```
INTEGER, PARAMETER :: dble=SELECTED_REAL_KIND(12 , 60)
REAL (KIND=dble) :: range , pressure
IF (dble<0) STOP ' Precision not available'
CALL rootsub(1.2_dble)
```

SELECTED_REAL_KIND returns the **KIND** that gives at least what you ask. (On a normal system this example would be 2, for standard double precision.) If your requirements are too demanding for any available scheme, a negative value is returned, and the **IF** statement checks for this.

9.2.3   Single precision versus double precision

In summary, the properties of IEEE FP numbers are:

Table 9.2   Single- and double-precision IEEE floating-point numbers

Quantity	Single precision	Double precision
exponent bits	8	11
binary fraction bits	23	52
xmax	3.402823E+38	1.7976931348623157D+308
xmin	1.175494E−38	2.2250738585072014D−308
eps	1.192093E−7	2.2204460492503131D−16

It has to be said that for scientific use both the accuracy and the range from 32-bit real numbers are borderline. six to seven digits of precision may well not be enough. You may want to work with numbers smaller than 10^{-38} or larger than 10^{38}: it is quite easy to get numbers of that magnitude in physics, as will be discussed in §9.3, but values above 10^{100} are scarce, and those exceeding 10^{308} are very unusual, even with a carelessly written program.

Using higher precision brings penalties which may or may not matter. Certainly your variables take up more space. (64 bits instead of 32, say.) For a handful of variables in a small program on a decent system, this doesn't matter, but if you're doing massive array calculations then it may. It may also slow you down: your compiler may provide standard and high-precision arithmetic, but it doesn't tell you that the standard arithmetic is done in fast hardware, perhaps by a floating-point co-processor, and the high-precision arithmetic is provided by a software library which is much much slower. Be on the lookout for some dramatic increases in processing time. But equally it may be that double precision is native to the hardware so there is no speed penalty.

One approach is to work in single precision unless you have reason to fear that this will not be good enough. Problems of loss of precision and calculations going out of range are avoided by careful algorithms, foreseeing possible problems. The opposite approach is to decide that you will always use double-precision variables unless you have reason to fear that will bring problems in speed or memory space. We tend to favour the latter: it is our opinion that *double precision is a wise choice for scientific computing*. This does not remove all possible problems, or remove the need for careful programming to avoid them, but it does make them less common.

C++ also defines **long double**; its size, in bytes, on any system is given by the operator **sizeof( )**, and is usually 10. Compare this with 8 bytes for a double and 4 bytes for a float. One implementation allows 19 SF and an exponent range of $10^{\pm 4932}$.

9.3   MULTIPLICATION AND DIVISION

When two floating-point numbers are multiplied the exponents are added, and the bias is compensated for. The fractional parts are multiplied; as these are both of the form $1.f_2$, the result must be more than 1 and less than 4: if it is greater than 2 then it must be shifted right 1 binary place, and the exponent incremented by 1.

The product of two binary numbers each with 24 significant bits is a number with 48 significant bits. So to store a completely accurate result of the multiplication of two floating-point numbers requires a word basically twice as long as the original. And indeed this is why the registers used for intermediate arithmetic are of extra length in many systems, and the origin of 'double precision'. But in fact, the loss of accuracy when you round the 48 bits down to 24 again by storing the answer in a 32-bit word is not usually serious. Multiplication (and division) don't generally give rise to devastating losses in accuracy in the same way that (as we shall see) addition and subtraction can. What's more of a worry are overflow and underflow, which occur when the exponents run out of their range.

9.3.1   Range problems for REAL and float variables

Exceeding the value $\simeq 3.4 \cdot 10^{+38}$ will cause the compiler to issue a fatal-error overflow message whereas underflows (progressive or strict – see §9.2.1) are usually quietly set to zero and not reported to the user. These can arise quite easily. Here is a simple calculation of the energy of a neutron beam.

Example 9.1   *Calculate the energy (in eV) of a neutron beam which has a wavelength of* $\lambda = 5 \cdot 10^{-6} m$.

Solution 9.1   *The formula is* $E = h^2/(2m\lambda^2)$ *where Planck's constant* $h = 6.626 \cdot 10^{-34} Js$ *and the neutron mass is* $m = 1.675 \cdot 10^{-27} kg$. *This is then to be divided by* $e = 1.602 \cdot 10^{-19} C$ *to convert the result into eV which are the natural units for the example. The result is* $5.242 \cdot 10^{-30} J$ *or* $0.3272 \cdot 10^{-10} eV$.

The initial coding (ES is exponential scientific, §8.2.2.1) might look like

```
{   /* ... */
    float  h , m , lam , q , E;
    h = 6.626e-34;  m = 1.675e-27;
    lam = 5.0e-6;   q = 1.602e-19;
    float h2  = pow(h , 2);
    float den = 2.0*m*pow(lam , 2);
    E = h2 / den;
    cout << " Energy in J and eV "
         << E << " " << E/q << endl;
}
```

```
!...
REAL  :: h , m , lam , q , E , h2 , den
h   = 6.626e-34    ! Planck
m   = 1.675e-27    ! neutron mass
lam = 5.0e-6       ! wavelength
q   = 1.602e-19    ! electron charge
h2=h*h ; den = 2.0*m*lam**2
E   = h2 / den
PRINT '(A , 2ES15.6)' , ' Energy for &
&neutron diffraction(J and eV) ', E , E/q
END
```

The results for E and $E_{ev} = E/q$ are both zero! The numerator, which has the value $h^2 \approx 4 \cdot 10^{-67}$ *underflows to zero*. Even though we multiply up subsequently the compiler has already set the result equal to zero and the damage has been done. Please note that these effects are compiler dependent. The problem here is easy to spot because of course the energy isn't really zero. But a correction term set to zero, due to underflow, in a more complex calculation may well go unnoticed.

The calculation attempt fails even more dramatically when the example is slightly changed to use $\lambda = 0.5 \cdot 10^{-10}$m instead, which is a wavelength suitable for the study of crystal structure. Then we obtain a *divide-by-zero* fatal error. Both the numerator $h^2 \approx 4 \cdot 10^{-67}$ and the denominator $2m\lambda^2 \approx 10^{-47}$ *underflow to zero*. A small amount of experimentation shows that reprogramming (on some compilers)

E = (0.5/m)∗pow(h , 2) / pow(lam , 2); | E = (0.5/m)∗h∗∗2 / lam∗∗2

avoids the underflow. However, this is *not* the way to do programming, where we are close to numerical boundaries and constantly must resort to tricks to circumvent them and it shows up the difficulty (which in this case is self-inflicted) of working the example with either the *wrong units*, or the *wrong precision*. Trivial, unnecessary and frustrating difficulties of this kind can be bypassed in this case at the outset. The remedy is to *work in physical units more suited to the example*, with wavelengths in nm, energies in eV, masses as their equivalent energy i.e. $mc^2 = 939.6$ MeV and $hc = 197.3(2\pi)$ eV nm instead. There is then no possibility of an underflow, and our program fragments become

```
// using appropriate units          ! underflow is avoided
float  hc , mc2 , lam , E;           REAL  :: hc , mc2 , lam , E
hc   = 1239.7;        // eV nm       hc   = 1239.7      ! eV nm
mc2  = 939.6e6;       // eV          mc2  = 939.6e6     ! eV
lam  = 5000.;         // nm          lam  = 5000.       ! nm
E = pow(hc/lam , 2) / (2.0∗mc2);     E = hc∗∗2 / (2.0∗mc2∗lam∗∗2)
cout << " Energy in eV " << E << endl;  PRINT '(A , ES15.6)' ,&
                                     &' Energy in eV ' , E
                                     END
```

Problem 9.5   *The simplest problem in atomic physics is to calculate the potential energy of the electron in the H-atom, for a radius $a_0 = 5.29 \cdot 10^{-10}$ m, called the Bohr radius. It is given by the formula*

$$E = -\frac{e^2}{8\pi\epsilon_0 a_0^2}$$

Write a program to calculate this: use various approaches and see how underflows arise and how they can be avoided. (ϵ_0 is $8.854 \cdot 10^{-12}$ in SI units.)

9.4 AVOIDING UNNECESSARY PRECISION IN CALCULATIONS

If many of the leading *guard digits* in the calculation are redundant, because they do not change, it is good practice to remove them before the computation and replace them at the end. Just as the previous section showed we should avoid unnecessarily large multiplicative factors by choosing sensible units, here we have to avoid unnecessarily large additive factors. The higher the accuracy of measurements, or the more complex the calculations, the more necessary this procedure is.

Example 9.2 *Measurements of the speed of sound in air yielded the results* 333, 336, 330, 329, 331 *m s*$^{-1}$. *Calculate the mean.*

Solution 9.2 *You could add the values up, and then divide by 5, and this is how you would program the task. The result* is* 331.8 *m s*$^{-1}$.

Or you can do it in your head by taking off a constant value of 330, and calculating $(3 + 6 + 0 - 1 + 1)/5 = 1.8$, and adding 330 to get the answer.

By working with reference to the value of 330 we not only make the sum easier for ourselves, and avoid the chances of mistakes (which are not considerations for the computer) but we also achieve an answer accurate to four significant figures by a calculation which only works with two. And this is a consideration that matters for computing problems.

The *false mean* is the formalisation of this common-sense idea. You choose an approximate value of the mean by inspection, and compute relative to it, correcting for the difference between the true and false means at the end.

Problem 9.6 *Four absorption peaks in the water-vapour spectrum are measured with a precise laser spectrometer. Their energies are* 12683.782, 12685.540, 12685.769, 12687.066 *cm*$^{-1}$. *Write a program to calculate the mean*† *using a naïve sum and a sum using* 12680.000 *as a false mean. Are the answers different?*

The formula expressing the mean, for N values labelled $k = 1, 2 \dots N$, is

$$\bar{y} = \frac{1}{N} \sum (y_k - m) + m \qquad (9.1)$$

where m is the false mean and $\bar{y}$ is the true one

$$\bar{y} = \frac{1}{N} \sum y_k = \frac{1}{N}(y_1 + y_2 + \dots + y_N) \qquad (9.2)$$

---

* Notice that the mean can be quoted to one additional figure of precision. Recall (e.g. *Barlow* §5.2.1) that the mean, $\bar{y}$, of N numbers each with *standard deviation* σ is $\bar{y} \pm \sigma/\sqrt{N}$.

$\dagger$ Notice the difference in concept between these two calculations. In the first we have five experimental values, with errors of probably ≈ 1 m s^{-1} and we seek the best mean value. In the second we are finding the mean, or centroid, of different values (the errors are ± 0.001 cm^{-1}) on the way to finding the best slope of a line which relates them.

9.5   PRECISION LOSS IN ADDITION AND SUBTRACTION

Consider what happens when two FP numbers are added or subtracted. If the exponents are not equal, then the smaller value must be adjusted downwards until its exponent matches the other.

These lower bits or digits are lost. There's no point trying to hold on to them, even if we could by using a double-length register, as the corresponding bits in the larger number are not specified.

The two significands are then added. It is possible that they give a value greater than 2 (but less than 4) and if so the exponent is incremented by 1 and the result is shifted down, thus losing another bit of significance. This is unlikely to be of importance.

What has more consequence is that *cancellation* can occur, either by adding two numbers of opposite sign or subtracting two numbers of the same sign. The exponent then has to be decremented, and the result has to be shifted upwards. Using decimals as a (comprehensible) example, $(1.23456 - 1.23455) \cdot 10^5 = 0.00001 \cdot 10^5 = 1.00000 \cdot 10^0$. This result has 6 significant figures on the face of it, but only the first digit is genuine. Such significance losses due to cancellations arise frequently. Expressions like $A - B$ where A and B are of similar size turn out to be distressingly common. It's vital to program in such a way as to avoid unneeded cancellations.

9.5.1   Avoiding cancellation losses: rewriting the algebra

Consider the difference $123.5^2 - 123.4^2$, which could have arisen, for example, from the common relativistic kinematics equation $m^2c^4 = E^2 - p^2c^2$. Evaluating this directly leads to $15252.25 - 15227.56 = 24.69$ where to achieve 4 SF(significant figures) in the answer, from 4 SF starting values, 7 SF were necessary in the calculation, the first three being guard digits. As shown in §9.3, a multiplication basically doubles the number of digits required for storage in order to retain the accuracy. If we used 4 SF in the calculation we lose most of the final accuracy and we would compute $15250 - 15230$ giving 20, not 24.69.

We *can* retain full accuracy by recognising that the expression *factorises* into

$$(123.5 - 123.4) \cdot (123.5 + 123.4) = 0.1 \cdot 246.9 = 24.69$$

The key is to *combine before multiplying*. This is used in Horner's method (§9.5.6).

Problem 9.7   *For a particle of momentum p MeV/c and mass m MeV/c²* *the velocity β can be found from the formula $\beta = p/\sqrt{m^2 + p^2}$ and then the momentum from $E = m/\sqrt{1 - \beta^2}$. Alternatively the energy can be found directly from $E = \sqrt{p^2 + m^2}$. Compare the results of the two formulae for a muon, of mass 105 MeV/c², for a range of momenta, such as 1, 20, 50, 100, 200 and 500 MeV/c. Discover where, and why, they disagree.*

Although the difficulties in these particular examples may be trivially overcome by using double-precision calculations, do not be lulled into a false sense of security if you elect to use a greater accuracy. Problems of this type will *always* be present in floating-point arithmetic of whatever length; they will be less frequent with the higher precisions but they still happen.

9.5.2   Rewriting the algebra: the exponential series

Suppose you need to calculate the negative exponential e^{-x}. All compilers provide the exponential function so at least we can check our answers! Also we shall achieve some insight into the possible problems and pitfalls concealed in the task.
You check the expansion and find:

$$e^{-x} = 1 - x + x^2/2! - x^3/3! + \ldots \qquad (9.3)$$

This converges rapidly for small x, but you know some mathematics and the fact that it converges for *all* x. So you program it. You require the value* for $x = 8$, let us suppose, so in it goes. Starting from $k = 1$ each term is $(-8/k)$ times the previous one. Your program calculates terms until the sum stabilises – i.e. when the next term does not affect the sum.

Disaster strikes, however, for the result isn't even close to the correct value, which is 0.000333546 . . .. It has the wrong sign(!) and is 10 times too large. What has gone wrong?

This is an example of the wrong choice of calculation leading to excessive cancellation. The terms increase to a magnitude of 416.1016 (for $k = 8$, 9) and then must cancel themselves exactly to allow the sum to decrease by 7 orders of magnitude, even to get the first digit of the result. But each term is only accurate to 7 figures at most so that all digits in the answer are worthless.

There has to be a better way! The first improvement is to *change the method* and convert the alternating sequence $(+ - + - + - \ldots$ to a positive one $(+ + + + + +...)$. This is easy: we calculate e^{+x} instead of e^{-x} and find the reciprocal at the end. Now at least the worst is over.

But if we want the result correct to 5 SF we also must take a great many terms in the sequence. After the 20th term (1.18 . . .) the result is only within a part in 1000 or so. Only after 80 terms is each term decreasing by a factor of 10 for each step in k. If we had *better convergence* then the calculation would be much more efficient. We achieve this by, once again, *changing the method*. Let us compute, instead, a suitable root of the number we want – say the 64th root. This means we obtain rapid convergence of $r = e^{+0.125} = e^{+8/64}$. Now to recover the result we need some multiplications. How many? We don't need 63 (fortunately), in fact just 6 are sufficient! Just follow a sequence of 6 successive squarings.

```
for (k=0; k<6; k++, r *= r);          DO k = 1 , 6
                                          r = r**2
                                      ENDDO
```

---

* This calculation is compiler-dependent. You may need $x = 10$ to reproduce this behaviour.

Please don't think of this as an artificial example. There are several important subtleties in its solution and all of these considerations, and more, are required in the design of the built-in functions provided by the libraries. It is of great importance to create reliable, robust and efficient software for the mathematics libraries so that scientific computing is possible. Yes, you have the benefit of their experience and are unlikely to need to calculate e^x yourself – but there are many other functions defined as the sums of series in this way, and you may well find yourself needing to calculate one of the ones that the compiler writers never got round to!

As far as mathematical functions go **Fortran** and **C++** provide only **SQRT, EXP, LOG, SIN, COS, TAN, SINH, COSH, TANH, ASIN, ACOS ATAN, ATAN2,** and **LOG10**. Some of these take a complex argument and/or produce a complex result. But if you need Bessel functions, the Gamma function, probability functions or the error function, for example, then you need to look elsewhere. A sensible place to start is *Numerical Recipes* (Press *et al* 1993a, 1993b, 1996) versions of which are available, with programs, for the **Fortran 90** and C languages.

9.5.3   Avoiding cancellation losses: the false mean

The *variance* of a set of N numbers $(y_1, y_2, \ldots , y_N)$ is*

$$s_y^2 = \frac{1}{N} \sum (y_k - \bar{y})^2 \tag{9.4}$$

where $\bar{y}$ is their mean, Equation (9.2). The variance may also be written

$$s_y^2 = \frac{1}{N} \sum y_k^2 - \left(\frac{1}{N} \sum y_k \right)^2 \equiv \overline{y^2} - \bar{y}^2 \tag{9.5}$$

Algebraically these equations, (9.4) and (9.5), are the same. Computationally they are different because $\overline{y^2}$ and $\bar{y}^2$ can be – and often are – very similar, and if so then Equation (9.5) gives cancellation losses. (Similar problems arise in calculating covariances : $\overline{xy} - \bar{x}\bar{y}$, and associated correlations. You have been warned.)

The advantage of Equation (9.5) over (9.4) is that you don't need to know the value of $\bar{y}$ to calculate s_y^2. This means you can do the whole calculation with one pass through the data, which may be useful for very large data sets. The false-mean technique can again be used, with the advantages of both (9.4) and (9.5). The variance is written as

$$s_y^2 = \frac{1}{N} \sum (y_k - m)^2 - (\bar{y} - m)^2 \tag{9.6}$$

This is true for any m, but if m is a good guess at the mean $\bar{y}$ then the second term is small compared to the first. In this formula the same deviations $(y_k - m)$ appear as in (9.1), and this time they are squared. So programming the mean and variance together saves effort. Note that (9.6) contains both (9.4), $m = \bar{y}$, and (9.5), $m = 0$.

---

* In some cases $N - 1$ may be appropriate in place of N. That's all we're going to say. This is not a statistics book.

The speed data of Example 9.2 give $(1/5)(9 + 36 + 0 + 1 + 1) = 9.40$ for the first term of Equation (9.6) and $(1.8 - 0.0)^2$ for the second term. The variance is thus $9.40 - 3.24 = 6.16 \text{ m}^2 \text{ s}^{-2}$.

Example 9.3 *Compute the variance of the water-vapour data using a false mean of, say, 12680.000. Compare the result (which is 1.368) with that obtained using Equation (9.5).*

Solution 9.3 *Our system gave the results in the following table:*

Table 9.3 Variance using different false means, with single-precision rounding errors

Value of m	$\frac{1}{N} \sum (y_k - m)^2$	$(\bar{y} - m)^2$	$s_y^2(SP)$	$s_y^2(DP)$
12685.529	1.368	0.000	1.368	1.368
12680.000	32.052	30.681	1.371	1.368
12600.000	469965.500	469963.813	1.688	1.368
0.000	160922912.000	160922896.000	16.000	1.368

The table shows the terms being subtracted in this simple example and the source of the loss of accuracy is obvious. We also show the results using double precision (SP means single-precision accuracy while DP means double precision) and it clearly solves all problems here – but how long would it do so if the number range became even smaller?

9.5.4 Averaging: the running mean

Suppose that you have measured 9 values and computed their mean value $\bar{y}_9$. Then you are inspired to make a 10th reading. Do you have to start all over again to calculate $\bar{y}_{10}$? By no means! From any set of values you can find the result for one more by *updating the mean*. The formula is rather simple for all we do is find the new excess $y_N - \bar{y}_{N-1}$ (the new point minus the old mean) and reduce its effect by dividing by the total number of points.

$$\delta_N = (y_N - \bar{y}_{N-1})/N \tag{9.7}$$

and update the mean accordingly:

$$\bar{y}_N = \bar{y}_{N-1} + \delta_N \tag{9.8}$$

(Expanding the right-hand side will show you that this is the same as $\bar{y}_N = \frac{1}{N} \sum y_i$.)

This approach is the basis of a much better method of calculating the mean value. This can then be used in a loop to find the mean of numbers stored in a vector. This works even with **m** set to zero on the first pass, or set to the first value, as is shown.

```
float mean(int n , float old , float y)          REAL FUNCTION mean(n , old , y)
{ //== returns mean, updated by y               INTEGER :: n
  //== y is the Nth value                        REAL      :: y , old
  return old + (y—old) / float(n);                 mean = old + (y—old) / REAL(n)
}                                                END FUNCTION mean
//== example of its use :                        !=== example of its use :
{                                                  REAL    :: m=0.0 , mean , &
  float m=0.0;                                   &speed(5) = (/1., 2.3, 4.1, 5., 6./)
  float speed[5]={10.,23.,41.,50.,60.};           INTEGER :: i
  for(int i=0;  i<5;  i++)                         DO i=1 , 5    ! averages speed
    m  =  mean(i+1 , m , speed[i]);                  m  =  mean(i , m , speed(i))
}                                                ENDDO
                                                 END
```

9.5.5   Updating the variance

The sample variance of the data, defined in Equation (9.4), can be updated in the same way (with δ_N given in Equation (9.7)):

$$s_N^2 = \frac{N-1}{N} s_{N-1}^2 + (N-1) \delta_N^2 \qquad (9.9)$$

You may wish instead to use the estimated *variance of the mean* itself, which is $\sigma_N^2 = s_N^2/(N-1)$:

$$\overline{\sigma}_N^2 = \frac{N-2}{N} \overline{\sigma}_{N-1}^2 + \delta_N^2 \qquad (9.10)$$

Equation (9.9) can be used from $N = 1$, while in Equation (9.10) set $\overline{\sigma}_1^2 = 0$ and start at $N = 2$. In both cases the variance at $N = 2$ is $\delta_2^2 = (y_1 - y_2)^2/4$.

Problem 9.8   *Write a function to compute both the mean and its standard deviation with just a single pass through the data.*

Such a function preserves accuracy, requires a single scan, and it can be computed as each data value arrives. By using a running mean one has the advantages of a false mean which automatically adjusts itself to the data.

This approach has another important advantage. The technique of floating-point addition means that when a small number is added to large one, the precision of the small one is lost. That's fair enough. We would accept that adding **1.E−1** to **1.E+6** would give **1.E+6**. But if you perform this addition a large number of times, say 10^6, you still get **1.E6**. And a loop which evaluates a running *total* may be doing just that

```
for(i=0;  i<n;  i++)              ! given REAL term(1 , N) , total
  total += term[i];                total = SUM(term)
```

As the loop progresses **total** gets ever bigger and bigger and the effect of **term** can get washed away. Use of the running mean avoids this trap.

9.5.6   Evaluating polynomials: Horner's method

Consider the computation of $(x-1.0)^4$ for $x = 1.01$: the result is clearly $1.000 \cdot 10^{-8}$. It can be expanded as a polynomial

$$(x - 1.0)^4 = x^4 - 4.0x^3 + 6.0x^2 - 4.0x + 1.0$$

The coefficients (part of the Pascal triangle) are similar in magnitude and the terms in the polynomial are all greater than 1, which means that (using single precision) no digits beyond 10^{-6} have any meaning. This can be shown with the following program fragments. (Intermediate results are stored to stop the compiler using double-precision intermediate values.)

```
float a , b , c , d , x , poly;
x = 1.01F;
a = pow(x , 4);  b = 4.0F*pow(x , 3);
c = 6.0F*pow(x , 2);  d = 4.0F*x;
poly = a − b + c − d + 1.0F;
cout << "polynomial " << poly << endl;
```

```
REAL :: a , b , c , d , x , poly
x = 1.01
a = x**4;       b = 4.0*x**3
c = 6.0*x**2;   d = 4.0*x
poly = a − b + c − d + 1.0
PRINT * , ' polynomial ' , poly
END
```

This gave us $-11.9 \cdot 10^{-8}$, not $+1.000 \cdot 10^8$! Printing the values for **a, b, c** and **d** shows that the massive leading cancellation $(1 - 4 + 6 - 4 + 1)$ denies to the calculation the precision needed in the 9th significant figure. Such calculational errors are unacceptable, for we tend to trust calculations and especially computers. After all, even a pocket calculator gets this right. But a pocket calculator typically works to 10 digits, not the 6 or 7 of standard floating point.

Alright, in this case we know that the polynomial is an expansion of a simpler form – but other polynomials are not! And they are going to have similar problems. How are we to compute such an expression accurately in (the normal) situation in which we do not know, at coding time, the factors of the polynomial? The answer is an elegant technique called *Horner's method* or *nested multiplication*. The principle is simply to *combine linear terms before multiplying* them, thus:

$$x^4 - 4.0x^3 + 6.0x^2 - 4.0x + 1.0 = [\{(x - 4.0) * x + 6.0\} * x - 4.0] * x + 1.0$$

At each step we multiply by **x** and add the next coefficient.

Problem 9.9   *Program Horner's method for this example. Then write a function to evaluate any polynomial, given a vector of coefficients.*

This works amazingly well in removing cancellation problems in polynomials. (Which is another reason put forward for not providing an easy-to-use exponentiation operator as part of the language.)

9.6 PRECISION IN PRACTICE

Many traps exist for programmers who forget that mathematics as practised on a computer is not the same as real mathematics. This is the theme of the whole chapter. Algebraic identities may not produce the same results computationally, due to the presence of rounding errors or of cancellation errors. The goal is, as always, *thoughtful programming* so that the surprises are fewer. There is no infallible, step-by-step guide that leads to correct programming. Important ingredients in the process include an open mind, an ability to foresee problems, critical thought and experience.

In this final section we highlight some numerical problems, and their solutions, to be found in the programming of the simplest tasks. Careful reading and understanding of this section will develop your abilities, if not to avoid problems altogether, then to recognise their cause when they arise in your own work.

Some of the difficulties were outlined in previous sections. Results typically differ by parts in 10^6 for a single-precision calculation, while at other times the failure is more catastrophic. Now we give some examples of what can happen.

9.6.1 Solving quadratic equations

Every book on programming treats this surprisingly rich problem: find the roots r_1, r_2 of the quadratic equation in the variable x with numerical coefficients a, b, c

$$f(x) = ax^2 + bx + c \equiv (x - r_1)(x - r_2) = 0 \tag{9.11}$$

which are given by

$$r_1, r_2 = (-b \pm \sqrt{b^2 - 4ac})/2a \tag{9.12}$$

The results, of course, depend on the value of the *discriminant** ($b^2 - 4ac$).

Table 9.4   Coefficients of the quadratic equation $ax^2 + bx + c = 0$ and the exact roots r_1, r_2

a	1.0	0.1	3.0	1.0	1.0	1.0	−1.0	3.0
b	5.0	0.5	−5.0	0.5	1000.001	−0.2	0.0	0.0
c	−6.0	−0.6	2.0	0.06	1.0	0.01	0.09	−0.27
r_1	2.0	2.0	1.0	0.2	−1000.0	0.1	0.3	0.3
r_2	3.0	3.0	2./3.	0.3	−0.001	0.1	−0.3	−0.3
case	1	2	3	4	5	6	7	8

Problem 9.10   *Write your own version of this code, or retrieve one you wrote earlier. Try the tests in Table 9.4, where the correct roots are also given.*

---

* Some books would have us believe that if $b^2 - 4ac < 0$ then there are no roots. Others only present test cases in which the roots are integers, and since all floating-point compilers treat integers exactly (up to 2^{18}) then no programming quirks can be revealed by such tests. Many example programs attempt to select $b^2 - 4ac = 0.0$ as a condition for equal roots, forgetting that reals are being compared (which cannot be done safely).

Example 9.4   *Consider the following questions: (i) All the roots arise from exact factors. How well do your answers agree with the true values? How well should they agree? (ii) What is the relationship of the roots in case (1) and case (2)? What is the error? (iii) How accurate is the small root in case (5)? How may this accuracy be preserved? (iv) Why are the roots in case (6) not equal computationally? How much accuracy is lost?*

Solution 9.4   *The errors that arise in Problem 9.10 are due to three principal causes (a) $b^2 \gg 4ac$, (b) that integers are not reals, and (c) $b^2 \simeq 4ac$. The discriminant, $b^2 - 4ac$, is large or close to zero, or is algebraically, but not computationally, zero. The rest of this section elaborates these points.*

If you program without very much thought then a blind translation of the algebra might produce something less than adequate:

```
//  Example of poor code            ! Not to be recommended either
{
    disc  = b*b − 4.0*a*c;          disc  = b*b − 4.0*a*c
    rootdisc = sqrt(disc);          rootdisc = SQRT(disc)
    r1 = 0.5*(−b − rootdisc) / a;   r1 = 0.5*(−b − rootdisc) / a
    r2 = 0.5*(−b + rootdisc) / a;   r2 = 0.5*(−b + rootdisc) / a
}
```

This is a fragment of a rough-and-ready program. The code should not be reached if a=0 (not a quadratic) or if **disc**< 0 (complex conjugate pair of roots). It illustrates *unnecessary* cancellations, since whatever the sign of **b** one of the roots loses accuracy. Should b^2 be large compared to $4ac$ then the two terms will almost cancel completely.
- **Large ratio between the roots :** $b^2 \gg 4ac$
  For cases like (5) where $r_1/r_2 \sim 10^7$, a loss of digits is inevitable using Equation (9.12). Since $4ac$ is small compared to b^2 the small root is

$$-b(1 - \sqrt{1 - 4ac/b^2}) \sim 10^3(1 - \sqrt{1 - 4.10^{-6}})$$

which can be expanded as a binomial and the large leading term cancelled. A much better way to program Equation (9.12) is to rewrite in terms of the large root (call it r_1) which does not lose accuracy through cancellation. Then we use the elementary fact that $r_1 \cdot r_2 = c/a$, and so

$$r_1 = -(b + sgn(b)\sqrt{b^2 - 4ac})/2a \qquad (9.13)$$

and $r_2 = c/(ar_1)$. The *sign* function $sgn(b)$ is $+1$ if $b \geq 0$, and -1 if $b < 0$. In **Fortran** this is **SIGN(a , b)** $=$ **ABS(a)***$sgn(b)$, with a=1.0, whereas in **C++** you make your own with **(b>=0.0)?+1.0:−1.0**.

- **Integers are different from reals**
  The roots of case (1) and case (2), i.e.

$$x^2 - 5.0x + 6.0 = 0.0$$

$$0.1x^2 - 0.5x + 0.6 = 0.0$$

are the same, 3.0, 2.0, but computing them results in 3.0000000, 2.0000000 and 2.9999995, 2.0000002, which, while being close, are different. These results are typical of all computers using floating-point systems.

- **Equal roots may be computationally different:** $b^2 \simeq 4ac$
  For case (6), above, the roots are equal but the simple program gives 0.10002... and 0.09998..., instead of 0.1000000 (7 SF). Here 3 SF have been lost since the errors are in the 4th or 5th significant figure. This is entirely due to careless, not to say casual, programming; we should check the computed value of *disc* and if it is computationally zero, within rounding errors, it should be *set equal to zero*. Since neither 0.2^2 or $4.0*0.01$ is exactly representable in a binary computer, their difference is likely to be within one bit of being zero and could be positive, negative or zero. Now the outcome of the program depends on the system programming of the *sqrt* function; on some compilers *sqrt*(0.0) is a fatal error while on others the result is set to zero. In some compilers *sqrt*($-x$) is a fatal error, in others it yields the complex number $0.0 + i\sqrt{x}$. The point here is a most important one: the *system* should not be making these decisions, it should be *you, the programmer*.

  The three sets [*a b c*] : case (6) [1.0 -0.2 0.01]; case (9) [1.0 -1.0 0.25]; and case (10) [1.0 -1.4 0.49] behave differently on compilers we have used. Case (6) has *disc* $\sim +10^{-8}$ so loses about 4 SF, case (9) has *disc* $\sim -10^{-7}$ so that *sqrt(disc)* fails, while case (10) has *disc* = 0.0 precisely. What does your compiler do with *sqrt*(0.0) ?
  Programming even the simple problems can be tricky. Vigilance is required.

9.6.2 Summing series

We will illustrate some aspects of computer calculations by summing various inverse powers of the natural numbers 1, 2, 3, 4, 5, ... Let us calculate three such sums, the inverse 4th power S_4, the inverse 2nd power S_2, and the inverse 1st power S_1. The first two results are known and are functions of π; in fact

$$S_4 = \frac{1}{1^4} + \frac{1}{2^4} + \frac{1}{3^4} + \frac{1}{4^4} + \frac{1}{5^4} + \cdots = \frac{\pi^4}{90} = 1.08232\ 32337\ 11138\ 191 \cdots \quad (9.14)$$

and

$$S_2 = \frac{1}{1^2} + \frac{1}{2^2} + \frac{1}{3^2} + \frac{1}{4^2} + \frac{1}{5^2} + \cdots = \frac{\pi^2}{6} = 1.64493\ 40668\ 48226\ 436 \cdots \quad (9.15)$$

The following code performs the sum for S_4:

```
//          program sum_4
#include <iostream.h>
#include <iomanip.h>
#include <math.h>
main()
{
  double pi_4=pow(M_PI , 4) / 90.0;
  double sum , one = 1.0;
  long int k , ktot;
  for( ; ; )
  {
    sum = 0.0;
    cout << "ktot = sum limit\n";
    cin >> ktot;
    if (ktot < 1) break;
    for (k=1; k<=ktot; k++)
      sum += one/pow(double(k) , 4);
    cout << setprecision(14) << sum
      << " " << pi_4-sum << endl;
  }
  return 0;
}
```

```
PROGRAM SUM_4
! Sums the inverse 4-th powers of
! the natural numbers up to ktot
! 1/1**4 + 1/2**4 + 1/3**4 ...
REAL(KIND=2)  :: sum , pi490
REAL(KIND=2)  :: one = 1.d0
CHARACTER  :: fmt*9
INTEGER  :: k , ktot
pi490 = (4.d0*ATAN(one))**4 / 90.d0
fmt = '(2F20.15)'
INF_LOOP: DO
  sum = 0.d0
  PRINT * , ' ktot=sum  of <= 10000'
  READ * , ktot
  IF( ktot < 1 ) EXIT INF_LOOP
  DO k = 1 , ktot
    sum = sum + one / REAL(k)**4
  ENDDO
  PRINT fmt , sum , pi490 - sum
ENDDO INF_LOOP
STOP 'SUM_4 concluded'
END PROGRAM SUM_4
```

By running these programs we can learn some interesting and initially surprising facts. First let us decide on precision – we shall try for 12 decimal places since the variables are stored to about 16 SF. This means that terms beyond **ktot = 1000** will be smaller than our target, although we might expect that sums of them may affect several of the digits in the last places. We obtain results for 10 terms, 100 terms, 1000 terms and 10000 terms. For k terms we list in Table 9.5 values of $\pi^4/90 - S_4(k)$, i.e. the amount by which the computed sum falls short of the limiting value. This increases, relative to k^4, as more terms are used! Nevertheless for 10^4 terms the error is only $3 \cdot 10^{-12}$.

Table 9.5   Summation of the series $S_4(k)$. k^{-4} is the size of last used term

terms k	sum $S_4(k)$	$\pi^4/90 - S_4(k)$	k^{-4}	err/k^4
10	1.08203 65834...	0.00028 6650...	0.00010 0000	3
100	1.08232 29053 44..	0.00000 03283 66..	0.00000 00100 00	33
1000	1.08232 32333 78304..	0.00000 00003 32834..	0.00000 00000 01000	330
10000	1.08232 32337 10804	0.00000 00000 00333 1	0.00000 00000 00000 1	3330

A small amount of reprogramming allows us to compute partial sums for S_2. The results are dramatically different. Table 9.6 shows that for 10^k terms the error is in the kth-place, i.e. 10^6 terms give us *only* 6-place accuracy!

Table 9.6   Summation of the series $S_2(k)$. k^{-2} is the size of last used term

k	sum $S_2(k)$	$\pi^2/6 - S_2(k)$	k^{-2}	error/k^2
10^3	1.64393 45666..	0.00099 95001..	0.00000 10000	10^3
10^4	1.64483 40718 48059..	0.00009 99950 00166..	0.00000 00100 000	10^4
10^5	1.64492 40668 98226 2..	0.00000 99999 50000 0..	0.00000 00001 000	10^5
10^6	1.64493 30668 51415 9..	0.00000 09999 99499 8..	0.00000 00000 010	10^6
exact	1.64493 40668 48226 4..			

Despite an enormous increase in computing effort we are getting nowhere! The error for 10^k terms is 10^{-k}, so that in this case of the sums of the inverse squares we can obtain a *computationally* accurate value to 12 decimals which is *mathematically* accurate to just 6 decimals. In fact it is easy to show that for n terms the error is less than $1/n$. From the computed differences we readily see that, more precisely,

$$S_2(k) \simeq \pi^2/6 - 10^{-k} + \frac{1}{2} \cdot 10^{-2k} \qquad (9.16)$$

What all this means is that because this sum has a painfully slow convergence rate we can only directly calculate 6 or 7 decimal places. The calculation has no appreciable rounding error; it just gives the wrong answer!

Finally let us tackle the last of our sums: that of S_1. It is the sum of the inverse powers:

$$S_1 = \frac{1}{1} + \frac{1}{2} + \frac{1}{3} + \frac{1}{4} + \frac{1}{5} + \cdots \qquad (9.17)$$

Again a trivial change to the code gives us the results, as in Table 9.7. To increase the speed we will work in single precision/float accuracy where about 10^{-7} is the relative accuracy. The totals rise slowly – after 10^3 terms, $S_1(10^3) \simeq 7.485\ldots$, $S_1(10^4) \simeq 9.788\ldots$, $S_1(10^5) \simeq 12.090\ldots$, $S_1(10^6) \simeq 14.393\ldots$, and after $2 \cdot 10^6$ terms, $S_1(2\cdot10^6) \simeq 15.086\ldots$. In fact we have nearly reached the limit of our patience and still we have no result. This is not surprising: the series S_1 is *divergent* and given enough terms the sum will exceed any given number e.g. 1000000000. There are two important points appearing here:

(1) Computing alone will not tell us whether a series is convergent or not,

(2) In this case *rounding error* is significantly affecting the results.

We establish the last point by a neat trick: we merely recalculate the sum in reverse order. The hugely different results are also in Table 9.7. When we start to think it becomes apparent that the 'natural' i.e. increasing order is the worst way to accumulate the sums since we are adding always smaller terms to a large partial result. If this result is accurate to, say, 7 decimals then as soon as we reach $k = 10^7$ then *all* of the contribution for that and *all* succeeding terms is lost in rounding error. This is unfortunate because a significant (not to say dominant!) part of the sum comes from this region. If we sum a million terms then all but the first one hundred thousand are less than 0.00001. By starting with these large terms we throw away 5 decimal places in the addition of all of the 900000 terms which are left. It turns out that adding terms from smallest to largest is the most efficient strategy!

Table 9.7   Summation of the series $S_1(k)$, with single-precision rounding errors

terms k	sum $S_1(k)$	reverse sum	difference	k^{-1}
10^3	7.48547 84...	7.48547 17...	0.00000 67...	0.001
10^4	9.78761 2...	9.78760 4...	0.00000 8...	0.0001
10^5	12.09085 ...	12.09015 ...	0.00006 9...	0.00001
10^6	14.35735 ...	14.39265 ...	0.03529 ...	0.00000 1
$2 \cdot 10^6$	15.31103 ...	15.08657 ...	0.22445 ...	0.00000 05

It is likely that the reverse sum is the more accurate for the reasons already given; a calculation in long variables gives $S_1(10^6) \simeq 14.39272\ 67228\ 6478 \cdots$ which seems to confirm our expectation.

9.7   EXAMPLE 1: LEAST-SQUARES FIT TO A STRAIGHT LINE

We now list the steps needed to obtain the best-fit line to a set of experimental data points. Some of the material will be familiar. We consider the case where there are N data points, labelled $x_k, y_k \pm \sigma, k = 1 \ldots N$. Each different x_k, which is known exactly, has a measured y_k, within a constant experimental error, σ.

Some data is given in Table 9.8 for measuring Young's modulus for a metal wire.

Table 9.8   Young's modulus measured data

mass kg	0.0	1.0	2.0	3.0	4.0	5.0
extension mm	0.9	4.2	9.8	14.5	17.3	21.9
error mm	0.5	0.5	0.5	0.5	0.5	0.5

Theory suggests a straight line

$$f = a + bx \tag{9.18}$$

and we wish to find the *best* values of a, b defined to be such that the sum of the squares of the residuals $y_k - f_k$ is a minimum (hence *least* squares). The details are given in many books, such as Barlow* (1989). In our case the value of the slope, b, contains Young's modulus, while the intercept, a, has no particular physical meaning.

---

* See the complete discussion in Chapter 6, §6.1 §6.2.

First we compute the variance and covariance of the data set:

$$var\_x = \frac{1}{N} \sum (x_k - \bar{x})^2 \tag{9.19}$$

$$cov\_xy = \frac{1}{N} \sum (x_k - \bar{x})(y_k - \bar{y}) \tag{9.20}$$

either by the updating formula, Equation (9.9), or directly. The best slope, b, is their ratio:

$$b = \frac{cov\_xy}{var\_x}, \tag{9.21}$$

while the best intercept, a, is the difference

$$a = \bar{y} - b\,\bar{x} \tag{9.22}$$

Next we find the theoretical values predicted by our values of a, b because we need the residuals to calculate the value of the quantity *chi-squared*:

$$\chi^2 = \sum \frac{(y_k - f_k)^2}{\sigma^2} \tag{9.23}$$

This should be about $N - 2$ if the scatter around the line is statistically consistent with the known experimental errors. The value f_k is best computed as $f_k = \bar{y} + b(x_k - \bar{x})$. We use the approximation

$$\sqrt{\chi^2} - \sqrt{N - 2} \simeq 0.0 \pm 0.7 \tag{9.24}$$

and agree that the fit is satisfactory if the result is within ± 2.1 (i.e. 3 standard deviations) of zero. Once this test is passed then errors on the parameters can be quoted, using the error on the mean, $\sigma_{\bar{y}} = \sigma/\sqrt{N}$ and the standard deviation of the x–values, $s_x = \sqrt{var\_x}$:

$$b \pm \sigma_b = b \pm \sigma_{\bar{y}}/s_x \tag{9.25}$$

$$a \pm \sigma_a = a \pm \sigma_{\bar{y}}\sqrt{1 + (\bar{x}/s_x)^2} \tag{9.26}$$

as in Table 9.9.

Table 9.9   Results for Young's modulus measured data

$a \pm \sigma_a$	$b \pm \sigma_b$	χ^2	$N - 2$	$\bar{y}$	$\bar{x}$
0.79 ± 0.36	4.26 ± 0.12	8.6	4	11.43 ± 0.20	2.500

Most of this is embodied in the following (partial) **C++** code.

```
int  LeastSquares(double x[ ] , double y[ ] , double sig , int n)
//==least—squares calculation  data  in arrays x[n] , y[n] , error  in sig
{                      //==========carry out checks for n > 2 , sig > 0.0 , etc
      double  x_bar , y_bar , var_x , var_y , cov_xy , a , b ;
      var_x  = var_p(x , &x_bar , n);
      var_y  = var_p(y , &y_bar , n);
      cov_xy = covxy(x , y , x_bar , y_bar , n);
      b = cov_xy / var_x;  a = y_bar − b*x_bar;
      double chisq = 0.0;
      for (int k=0; k<n; k++)
            chisq += pow( ( y[k] − y_bar − b*(x[k]−x_bar) )/sig , 2);
//========test for acceptable fit
      if (fabs(sqrt(chisq) − sqrt(n−2)) > 2.1)
          cout  <<  " Check data, fit is bad. Chisq = "  <<  chisq  <<  endl;
//========output results ...
          cout  <<  " intercept a = "  <<  a  <<  endl;
          cout  <<  " slope       b = "  <<  b  <<  endl;
//========output other details and graph the results. Plot residuals !
return  0;
}
double var_p(double p[ ] , double *p_bar , int n)
{        //========find variance of vector : updating variance & mean
    *p_bar = p[0] ; double varp = 0.0;
    for (int k=1; k<n; k++)
        {
            double dk = double(k+1);
            double del_k = (p[k] − *p_bar) / dk;
            *p_bar += del_k;
            varp = double(k)*(varp/dk + del_k*del_k);
        }
    return varp;
}
double covxy(double p[ ] , double q[ ] , double p_bar , double q_bar , int n)
{                    //========find covariance, using arrays & means
      double  covpq = 0.0;
      for (int k=0; k<n; k++)
          covpq += (p[k] − p_bar) * (q[k] − q_bar);
      return covpq / double(n);
}
```

Only the major details are here and the full program is necessarily more extensive. Topics such as the selection and reading of data files, the writing of results and summary files, the degree of user interaction, and so on, must be considered.

9.8 EXAMPLE 2: POLYNOMIAL FITS USING Fortran ARRAY FUNCTIONS

For the **Fortran** version we shall be more ambitious and fit a polynomial curve to the data with 2, 3, or 4 terms. We give only the briefest description and suggest consulting Barlow(1989), Rice(1983) and Press *et al* (1993a, 1993b). The equation is

$$y = a.1 + b.x + c.x^2 + d.x^3 + \cdots + h.x^{k-1} \tag{9.27}$$

where there are N measurements and hence all quantities except the coefficients are $N \times 1$ column vectors. The equations for $k = 3$, a quadratic fit, can be written as

$$Ap = y, \quad \text{or} \quad (1 \quad x \quad x^2)(a \quad b \quad c)^T = y \tag{9.28}$$

The column vector $p = (a \quad b \quad c)^T$ contains the 3 desired coefficients, which are the unknowns. It is generally unwise to fit more than $k = 4$ using simple polynomials. The idea is to create the *normal matrix* M, which is $A^T A$ and is 3×3 in this case. Such a matrix is symmetric (and positive definite). The normal equations are

$$Mp = A^T y \tag{9.29}$$

and the problem is to solve them for p. Gaussian elimination is a sound and reliable method, as is the one we choose – the Cholesky method. We decompose M into the product $M = LL^T$ in which L is lower triangular (zeros above the diagonal).

We write for our case (where the unspecified terms below the diagonal have the same values as those above it),

$$M = \begin{pmatrix} M_{11} & M_{12} & M_{13} \\ & M_{22} & M_{23} \\ & & M_{33} \end{pmatrix} = \begin{pmatrix} l_{11} & 0 & 0 \\ l_{21} & l_{22} & 0 \\ l_{31} & l_{32} & l_{33} \end{pmatrix} \begin{pmatrix} l_{11} & l_{21} & l_{31} \\ 0 & l_{22} & l_{32} \\ 0 & 0 & l_{33} \end{pmatrix} \tag{9.30}$$

It follows by equating terms that

$$l_{11} = \sqrt{M_{11}}, \quad l_{21} = M_{12}/l_{11}, \quad l_{31} = M_{13}/l_{11} \tag{9.31}$$

$$l_{22} = \sqrt{M_{22} - l_{21}^2}, \quad l_{32} = M_{23} - l_{21}l_{31} \tag{9.32}$$

and

$$l_{33} = \sqrt{M_{33} - l_{31}^2 - l_{32}^2} \tag{9.33}$$

For larger matrices the approach is similar. We next solve for an intermediate vector q from $Lq = A^T y = rhs$, working down the rows. The results are $q_1 = rhs_1/l_{11}$, $q_2 = (rhs_2 - l_{21}q_1)/l_{22}$, $q_3 = (rhs_3 - l_{31}q_1 - l_{32}q_2)/l_{33}$. Finally, we solve similarly for p using $q = L^T p$ and working up from the last row. The structure of L makes the solution straightforward. The programs implement this whole process. **Nmat = M** is the normal matrix.

```
PROGRAM   LS_fit_poly
IMPLICIT NONE
!   First construct the normal matrix = k x k from A = N x k
!   part of programs to solve least–squares polynomial fits to data
!   N = number of data points ,  k–1 = polynomial order   !   y = data at x
INTEGER :: i , j , k , N
PARAMETER(k=3 , N=8)        ! the user could read these in before allocation
REAL(KIND=2) , dimension(: , :) , allocatable :: A , Nmat , L
REAL(KIND=2) , dimension(:)     , allocatable :: x , y , rhs , p , q
ALLOCATE  ( A(N , k) , Nmat(k , k) , rhs(k) , x(N ) , y(N ) )
x = (/40, 80, 120, 160, 200, 240, 280, 320 /)          !==thermocouple deg C
y = (/0.38, 0.64, 0.84, 0.96, 0.98, 0.93, 0.80, 0.61/)  !==thermocouple voltage
x = x / 100.0 ! scale temperature vector
A(: , 1) = 1.0;    A(: , 2) = x;    A(: , 3) = x*x         !==build up A by columns
Nmat = MATMUL(TRANSPOSE(A) , A)
rhs    = MATMUL(TRANSPOSE(A) , y)
!   Next part of the program to  solve  least–squares polynomial fits to data
!   Form Cholesky lower triangle matrix L from normal matrix Nmat = A'A
ALLOCATE( L(k , k) ) ; L = 0.0      ! essential to zero the new array
DO  i = 1 , k
    L(i , i) = SQRT( Nmat(i , i) − DOT_PRODUCT(L(i , 1:i−1) , L(i , 1:i−1)) )
    L(i+1:k , i) = Nmat(i , i+1:k) − MATMUL( L(i+1:k , 1:i−1) , L(i , 1:i−1) )
    L(i+1:k , i) = L(i+1:k , i) / L(i , i)
ENDDO
ALLOCATE( q(k) ) ; q = 0.0         ! solve for q using Lq = rhs
DO j = 1 , k
    q(j) = ( rhs(j) − DOT_PRODUCT( L(j , 1:j−1) , q(1:j−1)) ) / L(j , j)
ENDDO
ALLOCATE( p(k) ) ; p = 0.0         ! solve for p using L'p = q
DO  j = k , 1 , −1
    p(j) = ( q(j) − DOT_PRODUCT( L(j+1:k , j) , p(j+1:k) ) ) / L(j , j)
ENDDO
PRINT * , ' solution to y = a + b*x + cx*x is a = ' , p(1) &
     &, ' b = ' , p(2) , ' c = ' , p(3)
DEALLOCATE( A , Nmat , L , q )
! ...
END PROGRAM   LS_fit_poly
```

Of course there is more work to do, so that we should not be too hasty in deallocating the arrays. The fit vector *fit* = Ap should be found, and then the residual vector, $y − fit$. The residuals are expected to scatter randomly about 0.0. It is useful to plot both the fit and the residuals. From the residuals, and the error(s) on the y-values, we then calculate χ^2 and compare with number of degrees of freedom, $N − k$, Equation (9.24).

A point which is easy to overlook when using the **Fortran** array output feature as in **PRINT** $*$, **B** is that the results are output in column order. That is, for example,

$$B = \begin{pmatrix} 1 & 2 & 3 \\ 4 & 5 & 6 \end{pmatrix}$$

will be output as 1 4 2 5 3 6. To see the mathematical form you need to use the transpose **PRINT** $*$, **TRANSPOSE(B)** which gives the order 1 2 3 4 5 6.

Normally one should test the argument before taking its square root but the construction of the matrix **A** guarantees that the arguments are positive.

Fortran is doing quite a lot of work in these **DO** loops and the result looks very like the algebra. This helps to reduce errors. To appreciate the difference study the **Fortran 77/C** versions (see e.g. references below) and note the triple **DO** loops, the special cases when $i = 1$ or $j = 1$, and the close attention to detail required. The ability of **Fortran** to skip the loop when the 2nd argument is less than the 1st tidies up the programming greatly and helps to clarify one's thinking.

A complete job on the fitting process would also compute the errors on the coefficients. Least-squares theory, e.g. Barlow (1989, Chapter 6.6), shows that these are proportional to the square roots of the diagonal elements of the inverse of the normal matrix. This is the *only stage* at which the inverse of a matrix should be calculated during the data-fitting process. Never compute it to find something else! There are several reasons for the prohibition. The polynomial system is well-known to be ill-conditioned, and many significant digits can be lost. This matters much less when only errors are at stake. Calculating an inverse is much more complicated than finding the coefficients directly, as we have done above, and is costly in operation count. The complete error matrix is needed when the errors on functions of the coefficients are needed. The inverse of $A^T A$ could be found as $(L^T)^{-1} L^{-1}$ but just the elements of L^{-1} are sufficient. All these details, and more, are to be found in Press *et al* (1993a, 1993b, §2.3, §2.9), and in Press *et al* (1996, pp 1038–9).

10

Object-oriented programming

After an introduction to the ideas of object-oriented programming, the ways of writing such programs are shown in both languages: an object's structure in terms of data and functions, and the way that operators specifically for your objects can be defined. The how and why of public and private data is explained. A more detailed look is given at some **C++** features such as inheritance and templates, which are missing in **Fortran**, and a simple program for list processing is given as an example.

Object-oriented programming is a technique for analysing problems and writing code. Its popularity has increased explosively in the past few years, and books and magazines are full of descriptions that tell you it's the greatest thing since the discovery of the flow diagram. It is certainly useful and important but, despite the hype, object-oriented programming is not a universal solution to all programming problems. It is at its best when used for large, complicated projects written by a team of people, as it provides a clear way of analysing a problem into component parts with clearly defined interfaces. These can be developed and tested separately, and readily adapted to changing requirements and circumstances, avoiding duplication of effort. For small programming problems it is unwieldy – a program to write 'Hello, world' in the object-oriented X windows system is many pages long. The object-oriented approach is not a substitute for thoughtful programming, but a tool – one of many – which the thoughtful programmer will use when appropriate.

Object-oriented ideas can be applied in any computer language, but some are specifically designed for them. C++ is one such, and the extensions made to the original C language are mostly to do with allowing the implementation of object-oriented ideas. The modules of **Fortran** are also readily used for defining objects.

10.1   WHAT IS AN OBJECT?

When we define an everyday object, be it a cat or a computer, then a full and useful definition specifies not just what it *is* but what it *does*.

Cat: A small furry domestic quadruped. Purrs, likes eating fish, and catches mice.

Computer: A collection of integrated circuits and other electronic hardware. Used for calculations, data management, and word processing.

Within programs we define various entities: integer and floating-point variables are the simplest, arrays and strings are slightly more complicated, and complete structures of arbitrary complexity can be defined, by using **TYPE** in **Fortran** and **struct** or **class** in **C++**.* So far so good, but these all describe what the object *is* and not what it *does*. To go beyond a mere *data structure* to an *object type* the actions that can be performed with it have to be specified. This is done by specifying a set of functions (and, in **Fortran**, subprograms) that can be used with it.

As an example, let's take 3-dimensional position vectors. The data definition just requires the specification of 3 floating-point numbers. But what do they do – or, more appropriately in this case, what can you do to them? For example:

- A vector can be rotated through an angle θ about some axis to give another vector.
- Two vectors can have their scalar product taken to give a scalar.
- Two vectors can be added to one another to give a third vector.
- A vector can be drawn on a screen, as a point or a line.

There are many more (subtraction, translation...). These will be done in the program by functions that you provide. Their specification is part of what you mean by a 'vector' in this situation. Which you decide to code will depend on the requirements of your particular case.

For another example consider a Monte Carlo program to simulate light production and detection in a scintillator. The objects are photons: each photon has a position, a direction, and an energy – that's the data. Photons may be produced, reflected, refracted, scattered, absorbed, and detected. Those are the possible operations.

> **Problem 10.1**   *In a statistical analysis you use histograms. Analyse these as objects: how could you store the data and what operations are needed?*

Such object definitions say *what* can be done but they do not say *how*. This provides a helpful way of analysing a problem by separating the *application* – the what – from the *implementation* – the how. The two can be treated separately. Suppose you are using your computer for a problem involving vectors – planetary orbits, perhaps, or molecular kinetics. In an object-oriented approach you first look at the application, and decide on the definitions of the relevant processes that can be done with and to a vector. You spend a considerable time on getting this right, referring to similar programs written by yourself and by others and consulting colleagues who have appropriate experience. You may find that there is a definition of 'vector' in an

---

* The difference will be discussed in §10.5.

existing *class library* that suits your requirements. Having settled these, part of the problem is concerned with applying the laws of dynamics using the processes defined for these objects, and the other part is the implementation of these processes. These can be decoupled and considered separately, and may well be done by two different people, or different teams of people. The object definition is the agreed interface where the two meet: the applications program can be written with no concern as to how the operations are performed, and the implementation program provides the operations with no concern as to what they are going to be used for.

Object-oriented programming gives the applications programmer the advantage that it's easy and natural to write programs using *top-down design*: to start by deciding *what* objects are wanted and (at the same time) what they will do: the details of how this is done are deferred. If a design is done well, you can phrase what you want to do naturally (this vector should now be rotated by 10 degrees about this axis) rather than in terms of detail (this one-dimensional array should now be multiplied by this two-dimensional array using a loop). Object-oriented analysis helps – perhaps even enforces – such sensible design.

Object-oriented programming helps the implementation programmer too. The list and descriptions of objects specified for the application will probably turn out to contain a lot of repetition, or near repetition. Using *inheritance* – which the **C++** language stresses, but **Fortran** ignores – you identify the common elements and then you don't need to repeat them. For example, in a large astronomy simulation you might handle stars, planets, moons, and comets. Each of these has a mass and a position; each can move through space. But a star has a luminosity and the others don't. Planets and comets have a parent star, and only moons have a parent planet. Planets, moons, and comets can orbit, but (single) stars don't. An object-oriented analysis of the problem will search out the *commonality* between these separate classes, and implement it as a *base* or *parent class* – say **body**, which has a mass and a position and for which the function **move** is defined – and then define the *derived* or *child classes* on this basis: a star is a body with a luminosity, and so on. This search for commonality is a vital part of the problem analysis, making it clearer what the problem really consists of, and resulting in less repetitive, shorter and simpler code.

The separation between the implementation and the application raises the matter of *data hiding*. If suitable functions have been defined then the application programmer should not need to know how the data is stored, and should not be allowed to access it. A vector can be stored as Cartesian x, y, z or spherical polar r, θ, ϕ coordinates, or in other ways. Which is chosen is a matter for the implementation programmer. (You might store both forms so as to save time spent converting from one to the other.) Provided the function definitions are maintained, the implementation programmer should be able to change the data definition without affecting the application programmer. This will be true if and only if the application programmer is not accessing the data. Indeed, if an ignorant application programmer starts writing to the data directly, he/she can introduce data inconsistency bugs. Both languages provide a system for hiding object data when desired. This will be discussed in §10.5.

10.1.1   Objects in C++

A *class* of objects is defined using **class** (or **struct**, see §10.5) which contains the definitions not only of the data, as was described in §3.9, but also of functions that act on such objects: it does this by specifying the relevant function prototypes.* This can all be done within one source file, but, except for small programs, for convenience and clarity the application program is best put in one file, the object definitions in a separate header file, and the code of the function bodies themselves in a third. So for a program to plot planetary orbits in two dimensions, defining 2D vectors as objects, we might have not only the main program file **planets.cpp** but a definition file **vec.h** and a file of functions **vec.cpp**. The header file might contain

```
// this is file vec.h — the object definitions
class vector                        // 2D vectors
  {                                 // no hidden data
    public:
      double x , y;                 // data — cartesian co—ordinates
      double  length();             // functions — length of a vector
      double  scalar_product(vector a);  //     scalar product with another vector
      void  rotate(double theta);   //       rotate this vector
      vector  scaled(double factor);  //       return a scaled copy
        /* ... */                   //       and so on
  };                                // remember the semicolon!
```

A **#include "vec.h"** in the application program makes the compiler read this file, and objects of this class can then be declared. The data within the objects – the *components* – can be accessed using the member-selection operator which is the full stop (or period) . and the syntax for functions is just the same, though the brackets make clear the difference between data like **mars.x** and functions like **mars.length()**.

```
// main program file planets.cpp
#include <iostream.h>
#include "vec.h"                    // read class definitions — note speech marks
main()
  {
    vector mars , venus , hale_bopp;  // declare some objects of class vector
    mars.x = 2.28e8; mars.y = 0.0;  // define position of planet mars
    mars.rotate(0.01);              // rotate 10 mrad
    cout << " Mars position " << mars.x << " " << mars.y << endl;
    cout << " Mars radius " << mars.length() << endl;
  }
```

The operations are provided by function code bodies, in a file **vec.cpp** which is separately compiled. These are needed when the program is linked into a complete executable, but not before.

---

* The actual function code bodies may be given, but it's usually best to give them separately.

```
// this is file vec.cpp  — the function bodies
#include "vec.h"          // need definitions in this file too
#include <math.h>
double vector::length()
{
    return sqrt(x*x + y*y);
}
double vector::scalar_product(vector a)
{
    return a.x*x + a.y*y;
}
void vector::rotate(double theta)
{        // rotate vector through theta radians
    double yy = y , xx = x;
    x = xx*cos(theta) − yy*sin(theta);
    y = xx*sin(theta) + yy*cos(theta);
    return;
}
vector vector::scaled(double f)
{        // return a scaled copy
    vector v;
    v.x = f*x;   v.y = f*y;
    return v;
}
```

There are two things to notice at once about these functions:

- The name of each function has the object class name incorporated, using the double-colon *scope resolution operator*. Thus the first function here is the **length** function that is referred to in the **vector** class definition, as opposed to any function of the same (fairly common!) name in any other class. The file may contain function bodies for several different object definitions.

- These functions apply to a specific object, so it is *not* given in the argument list. **length** finds the length of 'this object'. **scalar_product** evaluates the dot product of 'this object' with the vector given as an argument. The components of the object can be referred to within the function, without any member-selection operator.

Problem 10.2   *Write a member function (and header definition)* **rotate** *that (like the* **scaled** *function here) will return the rotated vector, but does not alter the original.*

Problem 10.3   *Write a class for 3D vectors, including the vector product.*

10.1.2   Objects in Fortran

In **Fortran** all the implementation goes into one **MODULE** file, which combines the jobs of the header and function files of C++. This is compiled first, which makes the information available for the subsequent compilation of an application. For example

```
MODULE vec                                  ! The module file vec.for
   TYPE vector                              ! define the data structure
      REAL :: x , y
   END TYPE vector
CONTAINS
   FUNCTION scalar_product(a , b)
      TYPE(vector) a , b
      REAL :: scalar_product
      scalar_product = a%x*b%x + a%y*b%y
   RETURN
   END FUNCTION scalar_product
   SUBROUTINE rotate_vec(v , theta)         ! Using a subroutine
      TYPE(vector) v
      REAL :: theta , temp_x
      temp_x = v%x                          ! save the x value
      v%x = v%x*cos(theta) − v%y*sin(theta)
      v%y = temp_x*sin(theta) + v%y*cos(theta)
   RETURN
   END SUBROUTINE rotate_vec
END MODULE vec
```

The **USE** statement in the application program pulls in this information, as in

```
USE vec
TYPE(vector) venus , mercury
   venus%x = 0.0
   venus%y = 1.0811e8
   CALL rotate_vec(venus , 0.010)
   ! ...
```

The **TYPE(vector) venus** is slightly clumsier than the C++ equivalent **vector mars;** but it does the same job of making the variable known and reserving two words of memory. Access to data components is similar except that the operator is '**%**' rather than '**.**'. The chief difference from C++ is that the functions are not specifically tied to the objects, so all arguments are given. That's why the scalar-product function has two listed arguments rather than the one of C++ and hence it would be invoked as **p = scalar_product(a , b)** rather than **p = a.scalar_product(b);**.

10.2   OPERATORS FOR OBJECTS

If you have defined a function which you would describe as 'adding' two of your objects, then you could call it **add** and write **x** = **add(y , z)** or (in C++) **x** = **y.add(z);**. But it would be more helpful and legible if you were able to write addition in the usual way, as **x** = **y+z**. This is so obviously desirable a feature that it is provided in both languages. You can also redefine subtraction, multiplication, or any other operator.* Relational operators are an obvious extension. (How would you define **a** > **b** if **a** and **b** are vectors? It depends on what you're using them for.)

10.2.1   Operator definitions in C++

In C++ you define a function **operator+** exactly as you would define **add**. When the compiler parses the expression **a+b** it decodes it as **a.operator+(b)**. Other operators such as − or < would be defined through functions **operator−** and **operator<**.

```
// ... part of vec.cpp
vector vector::operator+(vector b)
{
    vector temp;
    temp.x = x + b.x;
    temp.y = y + b.y;
    return temp;
}
```

```
//... part of applications program
vector delta;
delta.x = 100.0;
delta.y = 200.0;
mars = mars + delta; // perturb mars
```

The prototype **vector operator+(vector b);** must be in the header file **vec.h**.
Notice that you can't say **mars += delta;**, because += is an operator in its own right, and is not automatically decomposed into + and =. If you want to use such an expression then you have to define the function **operator+=**.

10.2.2   Operator definitions in Fortran

In **Fortran** the function is defined and given a name in the usual way. It is linked to the operator by an *interface block*. The module **vec.for** would include

```
! in the first part of the file
! before CONTAINS
    ! ...
INTERFACE OPERATOR (+)
    MODULE PROCEDURE vec_plus
END INTERFACE
    ! ...
```

```
! in the second part after CONTAINS
    FUNCTION vec_plus(a , b)
    TYPE(vector) :: a , b , vec_plus
    vec_plus%x = a%x + b%x
    vec_plus%y = a%y + b%y
    RETURN
    END FUNCTION vec_plus
```

---

* Except the triadic C++ ? : operator. And you can't change the precedence and association properties of an operator. a+b∗c will always be evaluated as a+(b∗c), whatever the actual operators + and ∗ do with these arguments.

The function itself is perfectly standard. You could use it as **vec_plus(a , b)** if you wished. Dyadic operators like this have two arguments, monadic operators have one. The clever bit is the **INTERFACE** description, which tells the compiler that when it meets an expression **a+b** it should check the types of **a** and **b**. Then, if they match the argument types of **vec_plus**, that function should be called to do the operation. If you have several classes of objects you can give a list of functions for the appropriate arguments. The **INTERFACE** can be given in the calling program or in the module itself: the latter is surely preferable. (It can also work by giving the prototype of an external function, but this is less relevant to the object-oriented approach where the operators and data are defined together in a module.)

In **Fortran** you can define your own operators, which begin and end with a full stop. For example if the vector product of two (3-dimensional) vectors is computed by a function **vec_prod** then with

```
INTERFACE OPERATOR (.cross.)
    MODULE PROCEDURE vec_ prod
END INTERFACE
```

you can write expressions like

```
TYPE(vector) :: a , b , c
    ! ...
a = b.cross.c
```

Problem 10.4   *Think about the implementation of the + operator for (a) integers (b) complex numbers (c) matrices and (d) strings.*

Problem 10.5   *Implement and test the addition and subtraction of vectors in your preferred language. Then define the ✦ operator for multiplication of a vector by a scalar and as the scalar product operator between two vectors.*

10.3   COMPARISON BETWEEN THE TWO LANGUAGES

The C++ class contains member functions, and it's clear that these functions belong to this class. In **Fortran** the link is not quite so direct. You use a **MODULE** invocation to access a particular module. This contains **TYPE** definitions. It also contains – presumably – subprograms that refer to and use the data type(s) defined at the start of the module. There is no specific link between the subprograms and the type definitions (except that they may share some private data). You could put the subprograms in a second module which refers to the types in the first, but that is needlessly complicated.

Although **Fortran** provides the ability to define new operators and C++ does not, in general there are many more object-oriented features in C++ than in **Fortran**. So in the remainder of this chapter we will sometimes restrict the discussion to what can be done in C++.

10.4   INITIALISATION, CONSTRUCTORS AND DESTRUCTORS

10.4.1   Initialising objects

A **class** declaration only tells the compiler that objects of this class *may* be created. Then when the compiler reads the statement **vector x;** it knows that the variable **x** will exist as a vector and will treat references in later code appropriately. At run time, when the program gets to the point corresponding to **vector x;**, the appropriate words of memory are reserved: the object **x** of the class **vector** is *constructed*. It stays in existence until the program leaves the block (marked out by braces) in which the declaration occurred. At this point the memory is available for reuse and whatever was put in it is lost. Things are slightly different if the variable is **static**; in this case it is constructed at the start of the program, and destroyed only at the end.

When a variable of a simple type is declared it may also be initialised: you can say **int n;** or **int n = 1;** as desired. This nice feature can be used for user-defined types too: if you provide an *initialiser* member function with the same name as the class it is called when the object is constructed. The class definition contains the prototype, for example **vector(double x0 , double y0);** (an initialiser function does *not* have a specified return type – not even **void**!) and the code is provided later:

```
vector::vector(double x0 , double y0)
{     // initialises a vector
    x = x0; y = y0;     // set values
}
```

You can then declare and initialise objects: **vector b(10.0 , 20.0);** declares the vector **b** to the compiler, and at run time it will allocate two words *and then* call your routine to initialise them. You can have several initialising routines for the same class if they have different argument lists (thanks to name-mangling; §5.2.7).

One less desirable feature of providing such an initialising function is that it then becomes compulsory! If the **vector(x0 , y0)** function is provided then a declaration which does not use it but simply says **vector c;** is illegal. Maybe you don't want to have to specify the initial values for every variable declaration: you can get round this by also providing a *default initialiser* with an empty argument list

```
vector();     // in the header     |     vector::vector() { }     // in function file
```

which will then be invoked by **vector c;**. The function body can be null, as here, or it could more sensibly be something like **x = 0.0; y = 0.0;**. The function brackets are *not* given in the declaration: **vector c();** is wrong. The compiler will treat this as a prototype declaration of a function **c** that takes no arguments and returns an object of type **vector**, which is what the syntax logically corresponds to. This is a rare appearance of a function call without brackets.

An initialisation, such as **vector c = b;**, where **b** is an existing object of the same type, copies each member (see also next section).

10.4.2 Constructor and destructor functions

This business becomes significantly more important when your objects contain pointers, because these point to an area of memory, which had better be something sensible. When a pointer variable is declared, space is automatically reserved for the pointer, but not for whatever it points to. It can point to an existing object, but it will often be used to point to a new object created with the operator **new** (which explicitly allocates memory and returns the pointer to it). So the initialisation is more significant, and is more appropriately known as a *constructor*.

For example, suppose you want a class for vectors of any size. The data will be stored in an array, and the object will have to know how big that is. The class definition will contain the data and the prototypes of the constructor and other functions.

```
// This is vector.h
class vector {      // general vector
    public:
    int dim;        // vector size
    double* data;   // pointer to elements
    vector(int n);  // constructor
    // ... other functions ...
```

```
// This is vector.cpp
vector::vector(int n);
{           // constructor body
    dim = n;
    data = new double(n);
}
```

Then a program can declare vectors of any size, such as **vector a(3);** or **vector b(3000);** or **vector* p_vec = new vector(100);**. When the declaration of the vector **a** is implemented at run time, two words will be allocated automatically for **a.dim** and the pointer **a.data**. Then the constructor routine that you supplied is invoked, which will initialise **dim** to 3, as specified, and then reserve space for the specified 3 doubles and set the pointer **data** to point to them. The members of the array can then be referred to as **a.data[0]** , **b.data[i]**, and so on.

Problem 10.6   *Alter the above initialiser so that the elements of the array are set to zero, or to a given value. You could use two functions, or one function with a default argument – which is better?*

When the program leaves the block in which an object was created the data components are freed. But if your constructor has specifically allocated data with **new,** that will stay allocated unless you remove it. You have to do this, or your memory will fill up with unwanted data. This is done by providing a *destructor function*: its name is the name of the object type preceded by a tilde ~ and it has no arguments. So you specify the prototype **void ~vector();** and the function body

```
vector::~vector()
    delete [ ] data;
```

Problem 10.7   *Why do you say* **delete [ ] data** *rather than* **delete data** *in the above example?*

Initialisation by copying pointer members is dangerous. You must provide a *copy constructor* to handle such cases: see the specialist literature for details.

10.5   PUBLIC AND PRIVATE DATA AND FUNCTIONS

For scientists writing their own programs it may make sense to have a complete 'freedom of information' policy for object data: it's a lot quicker to write **fred.x** than to define and use a function which only does the same thing, when you happen to know how the data is stored and that there are no plans to change this. We take the opposite line in this chapter, because data hiding is an important part of object-oriented programming. In an ideal object-oriented program object data is obtainable *only* through member functions. If an application needs the *x*-component of a vector then the implementer has written an **x_component** member function for this purpose. This will make the job of program maintenance simpler as there is less scope for ignorant users to create bugs – but you have to write a lot of member functions!

To make object data (and functions) accessible to the user in **C++** you specify them as **public**: to make them accessible only internally they are specified as **private**. Recognising that in some instances people will want their data to be mostly public, in others to be mostly private, there are in fact two ways of defining an object class: the keyword **class** or the keyword **struct**. In the former the default is private, in the latter the default is public. The following are thus completely equivalent:

struct fred
{
    double x , y , z;
    int q();
private:
    double a , b , c;
    int f();
}
	class fred
	{
	    double a , b , c;
	    int f();
	public:
	    double x , y , z;
	    int q();
	}

Data in **Fortran** can be declared as **PUBLIC** or **PRIVATE**, where **PRIVATE** means they can be used only within the module concerned. These can be specified for a whole module, for all or some elements of a type, and in a variable's attribute list, with the more detailed overriding the more universal. For example, with

```
MODULE graphics
PUBLIC                  ! data is public unless specified
   REAL , PRIVATE :: xscale
   REAL :: yscale
   TYPE vector
      PRIVATE
      REAL :: x
      REAL , PUBLIC :: y
   END TYPE vector
END MODULE graphics
```

a program with the statements **USE graphics** and **TYPE(vector) :: a** can refer to **yscale** and **a%y** but not **xscale** or **a%x**.

10.5.1   Friend functions

A C++ function can access an object's private data if the class definition declares it as a 'friend' of that class. This is done by putting the function prototype into the class definition preceded by the word **friend**. Such a function is called in the usual way, in contrast to the member function which is called for a particular object. Here are two ways of providing functions to find the length of a vector:

```
// header file
class vec
{
    double  x , y;
    friend double length(vec v);
};
```
```
// header file
class vec
{
    double  x , y;
    float length();
};
```

```
// function file
double length(vec v)
{
    return sqrt(v.x*v.x + v.y*v.y);
}
```
```
// function file
double  vec::length()
{
    return sqrt(x*x + y*y);
}
```

```
// application program
    vec a;
    /* ... */
    double x = length(a);
```
```
// application program
    vec a;
    /* ... */
    double x = a.length();
```

Problem 10.8   *Write two functions, one member function and one friend function, to calculate the scalar product of two vectors.*

One friend function you will want to write is the stream output operator << for your class. Its name is of course **operator**<<; it takes two arguments, a reference (§5.2.3) to an output stream (such as **cout** or **cerr**) and the object to be output (possibly as a reference to save copying large objects). It returns the reference to the output stream it was called with, so that statements like **cout** << **x** << **y;** do what you want.

So to print a vector as two numbers surrounded by brackets with a comma for separation, your class definition contains

```
        friend ostream& operator<<(ostream& out, vector& v);
```

and in the function bodies you define (notice that this doesn't contain the descoping operator!)

```
        ostream& operator<<(ostream& out , vector& v)
        {
            return out  <<  "("  <<  v.x  <<  ","  <<  v.y  <<  ")" ;
        }
```

★ 10.6   CONVERSIONS BETWEEN USER-DEFINED OBJECTS

You may also need to define type conversions between your objects, or between your objects and standard data types. For example, you might have a class **vec2D** of 2-D vectors and a more general class **vec** of n-dimensional vectors, and want to assign them to each other freely. In **Fortran** this is regarded as a redefinition of the assignment operator =, and an interface for the assignment operator is defined which is a subroutine of 2 arguments, in which the second is converted to the first. (The syntax **ASSIGNMENT(=)** mimics **OPERATOR(+)** but it is a little forced, as the operator = is the *only* assignment operator in **Fortran**!)

In C++ this conversion is regarded as a definition of the appropriate cast operator. The example below shows this.

```
// header file
class vec2D
{            // 2D vectors
    double x , y;
};
class vec
{            // General vectors
    int dimension;
    double* data;
    operator vec2D();  // cast to vec2D
    void vec(int n);   // constructor
    /* ... */
};

// function body file
vec::operator vec2D()
{
    vec2D v;
    v.x = data[0];
    v.y = data[1];
    return v;
}
```

```
TYPE vec2D
    REAL :: x , y
END TYPE vec2D
TYPE vec
    INTEGER :: N
    REAL :: DATA(N)
END TYPE vec

INTERFACE ASSIGNMENT(=)
  MODULE PROCEDURE vec2D_to_vec
END INTERFACE
    ! ...
SUBROUTINE vec2_to_vec(v,v2)
TYPE(vec) , INTENT = OUT :: v
TYPE(vec2D) , INTENT = IN :: v2
    v%data(1) = v2%x
    v%data(2) = v2%y
RETURN
END
```

Notice that the **C++** operator goes in the class which is being converted *from* and within that is the cast operator of the class being converted *to*. In the above examples, the **C++** program permits conversion from **vec** to **vec2D**, so an application program which has declared **vec2D v** as a two-dimensional vector and **vec w** as a general vector whose dimension happens to be 2 can give the assignment **w = v**. The **Fortran** example provides the opposite conversion.

Problem 10.9   *In your preferred language, write the other converter. Consider, in particular, general vectors whose dimension is not equal to 2.*

★ 10.7   INHERITANCE

Inheritance is a powerful way of avoiding code duplication which also helps clarify your thinking. You may have several classes in your problem which have a lot in common: for example suppose you're drawing moving coloured shapes on the screen. The shapes – circles, rectangles, squares ... – will be represented by objects of different classes – **circle rectangle square**. They have many properties that are different: a circle has a radius and a square doesn't. But they also have properties in common: each has a position on the screen and a colour. To avoid rewriting these many times, you group them into a common *base class* called **shape**, and then define further *derived classes* that differ in their different ways.

```
struct shape
    {
        double x , y;
        int colour_code;
        void move(float dx ,  float dy) {  x += dx;  y+= dy;  }
    };
struct circle : shape
    {
        double radius;
    };
struct square : shape
    {
        double side;
    };
struct rectangle: shape
    {
        double height ,  width;
    };
```

Then you can define objects for these classes (for example **square s;**) which will contain the specific derived-class quantities (**s.side**) and the base-class quantities (**s.x s.y s.colour_code**). Sometimes you actually define objects of the base class. Very often you don't.

The rules for public/private data access are detailed. Whether another program is allowed to access base-class data through a derived-class object – such as **s.x** etc. – depends on how the data is declared in the base class (if it's private then it's inaccessible) and on whether the derivation is public or private. If the derivation is private then base-class data can't be accessed, even if that data is public within the base class. The default depends whether the derived class is defined using **struct** (public) or **class** (private), and is overridden by the word **public** or **private** after the colon and before the base-class name.

Problem 10.10   *Rewrite the above using* **class**, *making all data public.*

There are some operations, like moving the shape position, which are so general that they are part of the base-class definition, and the prototype goes there. So if you have defined **square** s; and **circle** c; and assigned **s.x** etc, the positions can be changed through function calls like **s.move(10. , 3.);** or **c.move(x_step , y_step);** There are others which are peculiar to a specific derived class. For example you might have a function to interchange the width and height of a rectangle, which has no counterpart for circles and squares. These will be defined and used only for a particular derived class. (If they apply to more than one then you should set up another level of derivation.) But there are other operations which you will want to do generally, but are done in a different way for each derived type, for example scaling a shape by a factor. This means a separate member function for each derived class.

```
void circle::scale(double s)   { r *= s; }
void square::scale(double s)   { side *= s; }
void rectangle::scale(double s)   { width *= s; height *= s; }
```

s.scale(2.0) and **c.scale(2.0)** may look similar to us, but not to the compiler.

A initialiser/constructor and, if necessary, a destructor can be defined for the base class in the usual way. The derived-class constructor then must invoke the base-class constructor: the notation for this mimics the class definition. If we have

```
shape(double x_set , double y_set)          // initialisation for shape
{
    x = x_set; y = y_set;                   // initial position is given
    colour_code = 0;                        // initial colour 0
}
```

then the constructor for a circle would look like

```
circle(double x , double y , double r)      // initialisation for circle
    : shape(x , y)                          // uses shape initialiser
{
    radius = r;                             // and also sets radius
}
```

A derived class can have more than one base class, and a derived class can be used as a base class for the next generation. There is lots of scope for complication. Try and remember that the aim is simplicity and avoid doing things because the technology exists. But a thoughtful use of derived classes can be very powerful and helpful. Think about how you would code the above sort of situation without inheritance (as you would have to do in **Fortran**). You would declare a **shape** class, and declare objects in the class. The class definition would contain a key that specified explicitly whether this was a square or a circle or whatever. You could either put all possible components in the definition – a radius and a side and a height and a width – or include a couple of parameters whose meaning depended on the value of the key. The functions would be full of **case/switch** statements based on the key. All very messy.

10.8 POINTERS AND OBJECTS

As your programs and your programming get more sophisticated, you will find that pointers are used more and more. This is natural. Objects have – speaking loosely – connections with other objects. In an astronomy program you may have moons (each one object) each belonging to a particular planet (another object). An accounting program will link orders to stock items. Graphic shapes may be linked to an object describing the device on which they are displayed. Data measurements can be linked to their experiment, and to a fitting curve.

You can maintain such data in each object, but this is inelegant and dangerous. You could include in your definition of class **moon** items giving the name and mass of the parent planet, but it is easier to define each planet just once, with a **planet** class, and give a pointer from each moon object to the planet. It is also a lot safer! Values may change. (All right, the mass and name of Jupiter are unlikely to change. But the position does.) If there are multiple copies of these values you have to be sure you update all of them – miss any out and the integrity of your data is lost.

These links may just be for reading data (and **const** should be freely used) or they may involve writing it. A moon will not affect a planet (short of falling into it) but when an order is dispatched the level of the appropriate stock item gets depleted.

Problem 10.11   *Define a class* **moon**, *using both methods of including the planet information, and use it for the moons of Mars. If you are not convinced of the merits of pointers, repeat this for the moons of Jupiter.*

Pointers are also more adaptable and flexible. Having declared the class **planet** you may define them individually, or more neatly with an array.

```
class planet                         class planet
{                                    {
public:                              public:
   char* name;                          char* name;
};                                   };
planet mercury , venus , ... pluto;  planet p[9];
   strcpy(mercury.name , "Mercury");    strcpy(p[0].name , "Mercury");
   strcpy(venus.name , "Venus");        strcpy(p[1].name , "Venus");
   /* ... */                            /* ... */
   strcpy(pluto.name , "Pluto");        strcpy(p[8].name , "Pluto");
```

Or you can work at run time, reading the names from a file:

```
planet p[9];    for (int i=0; i<9; i++) { cin >> p[i].name; }
```

Array names are pointers and the loop can be written yet more neatly as

```
for (int i=0; i<9; p++ , i++) { cin >> p–>name; }
```

and the operation in the loop body, and similar manipulations, work whether the pointer is (as here) derived from an array name, or from a linked list (which would be more adaptable), or from a pointer within an object of another class, or wherever.

A helpful feature included in **C++** to make pointer usage easy is **this**, which is a pointer within a class definition pointing to the object itself. So in a member function of the **planet** you can refer to **mass** or **this—>mass**. This can be useful if, within a member function, you want to call a non-member function for this object.

One elegant use of **this** is with a set of routines which perform various operations and then return the object that called them. For example, if for the **vector** class the member functions **scale(double s)**, **rotate(double theta)**, and **translate(double stepx , double stepy)** have been defined, and all return an object of type vector, finishing with **return *this;**, then the sequence of operations on some vector **arrow**

  arrow.scale(2.0);   arrow.rotate(45.0);   arrow.shift(−10. , 10.);

can be written

  arrow.scale(2.0).rotate(45.0).shift(−10. , 10.);

Problem 10.12   *Show this. In particular, is the order correct?*

There can be pointers to derived classes, and there is nothing special about these. There can also be pointers to the base class. These are more interesting and flexible because a pointer to the base class can also point to any class derived from that base. If you have a base class **shape** and derived classes **circle** and **square**, and have declared **c** as a circle and **s** as a square, then you might also declare

  shape* p_shape;
  circle* p_circle;
  square* p_square;

Then it is legal to say **p_square = s; p_circle = c;** and illegal to say either **p_square = c;** or **p_circle = s;**, whereas **p_shape = s;** and **p_shape = c;** are both legal. This can be useful: you can form lists of shapes, using a simple linked list, where shapes of different sorts are strung together in a general sequence. You can pass a shape to a function which will manipulate it without worrying about which shape it is.

This is fine provided the function doesn't need to do anything specific. Suppose you have **draw** functions defined for the square and the circle, but not for the general shape as that can't be defined. You want to say **p_shape—>draw()** because when it is called at run time, **p_shape** will be pointing to a square or a circle. But with the set-up as given, the compiler will say that you are trying to apply a member function **draw** to an object of class **shape**. No such function is defined, so the compiler will object.

To avoid this you can provide a function prototype for **scale** as part of the **shape** definition, using the keyword **virtual**. There is no corresponding definition because this function is never called as such. What happens is that the compiler is happy because it has such a definition: it also reserves a word (typically) in each object of this base class to specify which of the possible derived classes it belongs to. Then at run time this is used as a key to invoke the member function of the correct derived class. (If you do provide a definition for a **virtual** function then it will be called as a default if no function of this name has been defined for the derived type.)

⋆ 10.9   TEMPLATES

Templates are another way of generalising classes. You use one to define several classes which have the same structure, the only difference being the types of their components. For example, in §10.4.2 we defined a class for vectors of arbitrary dimension. Their components were stored as type **double**, but you might want some vectors stored as type **float**. You could duplicate the entire class definition replacing **double** by **float**, and giving it a different name, but that would be tedious (all the member functions would need defining twice as well!). Instead you can use a template. The class definition is preceded by the keyword **template** and then a parameter list in angular brackets. Angular < > brackets are the characteristic signature of templates. Thus

```
// This is the header file vecs.h
template <class F> class vec
{              // general vector
  public:
    int dim;      // vector size
    F* data;      // pointer to elements
    vec(int n);   // constructor
    void zero();  // zero the vector
      /* ... */   // other functions...
};
```

```
// This is the file vecs.cpp
template <class F> void vec<F>::zero()
{        // defines the function zero()
    for(int i=0; i<dim; i++) data[i]=0.0;
}
template <class F> vec<F>::vec(int n)
{        // constructor function body
    dim = n;
    data = new F(n);
}
```

Then you can declare and construct such vectors by **vec<double> v1(100)**; or **vec<float> v2(100000)**; or use any other available type for which **data[i]=0.0** is legal. Notice that the return value type still comes just before the function name, and that the parameter appears in the scope name but not the function names.

What happens is that you write the *template* for **vec**. When you specify an actual object for this template with a specific parameter, the compiler *instantiates* the template to get the class. The *classes* **vec<float>** and **vec<double>** are just like normal classes except they're not specifically written down – though you can see what they are by substituting the argument for the parameter in your template. The member functions are template functions (§5.3.1).

You can give more than one type in the argument list – for instance you might sometimes want to store the dimension in something other than an ordinary **int**:

```
template <class I , class F> class vect
{           // general vector
  public:
    I dim;        // vector size
    F* data;      // pointer to elements
      /* ... as before */
};
```

```
template <class I , class F>
vect<I , F>::vect(int n)
{        // constructor function body
    dim = n;
    data = new F(n);
}
```

And then say, for example **vect<char , float> v3(127)**;.

You can substitute other sorts of parameters. The word **class** specifies that what follows is the name of a class, and the term is extended to include the standard system types. Other parameters similarly need to specify what they are, for the benefit of the compiler when it checks the syntax. Here is another way of defining vectors of arbitrary length:

template<int N> class vector { double data[N]; };

And then you can define **vector<2> a , b , c; , vector<4> d , e , f;** and so on.

10.10   AN EXAMPLE: THE LIST

A list is a flexible data structure that lends itself nicely to object-oriented design and the use of **C++** pointers. A simple list contains a number of elements known as *links* or *items*, each of which has some data content, and also a pointer to the next element (this pointer is zero for the last element in the list). More complicated lists can also have pointers to the previous element; this makes many operations easier, but you have to make very sure that there is no possibility of forward and backward pointers accidentally becoming contradictory!

Suppose you have some data on stars, giving their names, magnitudes, and distance (in light years) in a file which is easily readable, from a text file, say, **star.dat**:

```
Sirius        −1.50    8.6
Canopus       −0.73    98.0
Alpha−Centauri   0.10    4.3
Vega     0.04    26.0
Arcturus    0.0    36.0
Capella    0.05    45
Rigel     0.08    600
Procyon    0.34    11.4
   ...
```

We will store this data as a list rather than 3 arrays as (a) we don't know how long the file is and (b) we are going to perform various operations, such as sorting the data, which are typically best implemented using lists.

Knuth(1968, p. 235) lists 9 actions which one should be able to do with a list:

1 Access the kth item
2 Insert a new item
3 Delete an item
4 Combine two lists into one
5 Split one list into two
6 Copy a list
7 Count the number of items in a list
8 Sort a list
9 Search a list

Specifying all these gets us a long way towards specifying the list as an 'object'.

Below is the header file containing three class definitions. First the actual information about stars. The data members are obvious, and you can't do much with a star so no member functions are necessary. We also declare friends that need to access the private data: input and output functions, and functions in the class **list** because the list manipulations involve doing things with star data.

The *links* that form a chain are defined next. Each comprises just a data object and a pointer to the next link. Again there are no member functions, but some friends.

The list itself contains only one data item – the pointer to the first link. But to compensate there are a large number of functions; they correspond to Knuth's list above. The argument lists and return types show a bewildering display of asterisks and ampersands. Let's work through them.

Access to the kth item is provided through the square-bracket operator. Given a list **starlist** then **starlist.operator[ ](n)** is equivalent to the familiar shape of **starlist[n]**. So it looks just like an array. What it returns, in this case, is a pointer to the link: it would have been equally possible to return a reference to the link, or the data element in the link. This was a matter of choice. The advantage of doing it this way is that if the element doesn't exist you can return a null pointer.

insert will create a new link after the existing link **n** – which may be zero for insertion at the beginning, and copy the data from **∗s**. Using a pointer means this can readily be made an optional argument: you might want to insert an empty link which will then be filled.

chop is given a link pointer and deletes the next link in the chain, after locating the link after that. Awkward, in that the link *before* the chopped one must be found.

merge is passed a list as an argument and adds it on to the last member of this particular list object. The argument is passed by reference as it is then set to a null list, to prevent links being members of two lists simultaneously.

tail splits this list object at the link whose pointer is given as an argument, and returns a pointer to a new list that is created for this purpose.

copy also creates a new list, but this time it is returned as an argument (hence the ampersand to denote a reference).

length returns the number of links in this chain. You could argue that this is wasteful and it would be better to keep the length as a data member. The problem with this is that you have to be really sure that all the operations do update the length correctly. (Suppose you increment the length by 1 and then add a new link – but creating the new link fails because you ran out of memory!) And even if your programming is perfect, suppose you update the length and change the list correctly, but in between those actions another process running in parallel and sharing the same data tries to use it. Storing the length is redundant information, and experienced programmers avoid it unless it's really necessary.

sort has no arguments or return value, it just rearranges the list object.

Searching is provided by another square-bracket operator. The difference between this and the first one is in the argument list: this one will search for a link with a star of the given name. So you can say **starlist["Sirius"]** as easily as **starlist[97]**.

The final function is used by the others: it finds the end of the list, the null pointer which will be used when another link is appended. The odd return type – pointer to pointer to link – is needed to cater for null lists when this is the list **root** pointer, and non-null lists in which it is the **next** pointer of the last link. Both of these are pointer-to-link, and as we want to write to them it returns a pointer to this location.

```
// this is the header file mylist.h

class star
{
    char name[20];                  // data
    float magnitude;
    float distance;
                                    // functions declared as friends
    friend ostream& operator<<(ostream& out , star& s);
    friend istream& operator>> (istream& in , star& s);
    friend class list;              // list functions need to know about stars
};

class  link
{    // only 2 items:
    star data;                      // a data item
    link *next;                     // pointer to the next link
    friend class list;              // list functions need to know about links
    friend ostream& operator<<(ostream& out , link& ell);
};

class  list
{
    link* root;                     // points to first link – this is the only data  item!
  public:
    link* operator[ ](int n);
    void insert(int n , star* s=0);
    void chop(link*l);
    void merge(list& add);
    list* tail(link*l);
    void copy(list* &n);
    int length();
    void sort();
    link* operator[ ](char* c);
    link** final();
    friend ostream& operator  << (ostream& out , list& s);
};
```

Here is a typical application program using these objects. Data is read from the file by directing it to standard input so the star data is read using the **operator>>** function. So if **starexe** is the executable file we could use the command **starexe < star.dat**

```cpp
#include <iostream.h>
#include <iomanip.h>
#include <string.h>
#include "mylist.h"        // on the previous page

main()
{
    list starlist;
    star s;
                                    // read in the list and print it out
    while(cin >> s) starlist.insert(starlist.length() , &s);

    cout << starlist.length() << " items in list" << endl;
    cout << " original  list\n" << starlist << endl;


                                    // split list after 3rd member
    link* ell = starlist[3];
    list* tlist = starlist.tail(ell);
    cout << " after split\n" << starlist << endl;
    cout << *tlist << endl;
                                    // make a copy
    list* newlist;
    starlist.copy(newlist);
    cout << " copy \n";
    cout << *newlist << endl;
                                    // delete the entry after Rigel
    ell = (*tlist)["Rigel"];
    tlist->chop(ell);
    cout << " deletion \n";
    cout << *tlist << endl;
                                    // merge the two lists again
    cout << " merge \n";
    tlist->merge(starlist);
    cout << *tlist << endl;
                                    // sort the list
    cout << " sort\n";
    tlist->sort();
    cout << *tlist << endl;
}
```

Now we consider the functions. The input and output are straightforward. Star data is output in a neat format, a link is output as the star data and then either an arrow or **END** to show if another link follows. A list is output as a series of links.

```
istream& operator>>(istream& in, star& s)
{
    in >>s.name >>s.magnitude >> s.distance;
    return in;
}
ostream& operator<<
    (ostream& out , star& s)
{
    out << setw(20) << s.name
        << setprecision(4) << setw(8)
        << s.magnitude
        << setprecision(4) << setw(9)
        << s.distance;
    return out;
}
```

```
ostream& operator<<
    (ostream& out , link& k)
{
    out << k.data <<
        (k.next? "— —>":" END") << endl;
    return out;
}
ostream& operator<<
    (ostream& out , list& u)
{
    for(int i=1; i<=u.length(); i++)
    out << setw(5) << i <<  " " << *u[i];
    return out;
}
```

The **length** function just goes through the list and counts. The **final** function is similar in form but returns something different. Notice two different ways of running along the list. Both test a pointer using the fact that this will return 'false' for the value zero. The indexed access does a similar walk-through.

```
int  list::length()
    {                    // length of list
        int j;
        link* k = root;
        for (j=0 ;  k ;  j++ , k=k—>next);
        return j;
    }
link** list::final()
    {                    //  get final element
        link**p = &root;
        while(*p)   p = &(**p).next;
        return p;
    }
link* list::operator[ ](int n)
    {                    //  access by index
        link* k = root;
        if(k == 0) return 0;        // case if list is empty
        for(int j=1 ; j<n ; k=k—>next , j++) if(k—>next == 0) return 0;
        return  k;
    }
```

Then there are the more involved manipulations. The code is intended to be self-explanatory (as always!).

```
list* list::tail(link* k)
{        // chop off the tail after link k
list* p = newlist;      // to hold tail
p->root = k->next; // transfer
k->next = NULL;    // list drops tail
return p;           // return the tail
}


void   list::chop(link* l)
{        // remove one link
link* t = l->next;    // link for deletion
l->next = t->next;  // skip over
delete t;       // remove dead  link
}


void  list::copy(list* &n)
{        // copy this list to new one
n = new (list);   // create new list
link* p = root;   // point to 1st link
if( !p ) { n->root = NULL; return; }
link** q = &(n->root); // for new list
do
{
  *q = new(link); // make a newlink
  (**q).data = p->data; // copy data
  p = p->next;       // nextlink to copy
  q = &(**q).next; // where to copy to
} while (p);          // repeat till NULL
(*q) = NULL;      // terminate new list
return;
}
```

```
void list::insert(int n , star* s)
{        // insert new link after link n
link**l = &root;
for(int j=1; j<=n; j++)l = &(**l).next;
link*t = *l;      // remember this link
*l = new link;  // make a new link
if(s) (**l).data = *s;//copy data if present
(**l).next = t; // next link set to old value
}


void  list::merge(list& add)
{
  link** ell = final(); // go to end of list
  *ell = add.root;    // end of old list
                 // points to start of new list
  add.root = NULL; // set other
                 // list pointer to null
}


link* list::operator[ ](char *c)
{        // access by name
  link* l = root;
  while(strcmp(c , l->data.name))
  {
    l = l->next;
    if( !l ) return NULL;
  }
  return l;
}
```

And finally the sort function. There are various sorting techniques. The most basic is to search through n elements for the first, then through the remaining $n - 1$ for the second, and so on. This involves of the order of $n^2/2$ comparisons. A better way (which, incidentally, works well when you have a pile of items to sort in the real world) is to split the list in two, sort each separately, and then merge them (i.e. repeatedly take the lower of the two items on top of the two sorted sublists). This involves only of the order of $n \ln_2 n$ comparisons.

```
void list::sort()
  {
    int n = length();                      // get length of this list
    if(n == 1) return;                     // if length is 1 list is sorted
    list* a = this;                        // pointer a points to this list
    link* ll = (*a)[n/2];                  // find link halfway down
    list* b = a–>tail(ll);                 // split list in two
    a–>sort();                             // sort them both
    b–>sort();                             //  (this is a recursive call)
    list *nn = new list;                   // nn used for the merged list
    link** q = &nn–>root;                  // *q points to end of merged list
    link** r = &a–>root;                   // *r is used to run along list a
    link** s = &b–>root;                   // *s does the same for list b
    while(*s != 0 || *r != = 0)            // merge till end of both a and b
  {
    if ( *r == 0 || ( *s != 0  &&
    (**r).data.distance  >  (**s).data.distance) )
      {
          *q = *s;                         // take the item from list b
          q = &(**s).next;                 // next location on merged list
          *s = (**s).next;                 // move one down list b
      }
    else
      {
                                           // take item from list a
          *q = *r;                         // take the item
          q = &(**r).next;                 // next location on merge♡list
          *r = (**r).next;                 // move one down list a
      }
  }
    this–>root = nn–>root;                  // merged list pointer to list root
    return;
  }
```

The next page gives the output which is produced.

Problem 10.13   *Adapt the above to lists of library books (author, title, date, cost and anything else relevant).*

Problem 10.14   *Adapt the above to lists of anything using template classes.*

Knuth actually specifies that one should be able to sort and search a list on *any* field. The above can easily be adapted to sort on name or magnitude rather than distance, but how could this be done generally, so that the application could sort on magnitude or distance as it chose? One way is to add a messy extra argument.

Another way is to add a messy global variable. The clean way would be to define derived classes **star–by–name**, **star–by–distance** and **star–by–magnitude** which inherit everything from the base class **star** except for the function **operator>**, which is defined separately for each of them using the appropriate data component. The sort function becomes a template function with a class specified; it casts whatever it's given into that class, and the correct relational operator is then applied. We leave this as an exercise for the student.

Problem 10.15   *Do it!*

The output from the list program of §10.10, using **star.dat**, is

8 items in list

original list

1	Sirius	−1.5	8.6– –>
2	Canopus	−0.73	98– –>
3	Alpha–Centauri	0.1	4.3– –>
4	Vega	0.04	26– –>
5	Arcturus	0	36– –>
6	Capella	0.05	45– –>
7	Rigel	0.08	600– –>
8	Procyon	0.34	11.4 END

after split

1	Sirius	−1.5	8.6– –>
2	Canopus	−0.73	98– –>
3	Alpha–Centauri	0.1	4.3 END

1	Vega	0.04	26– –>
2	Arcturus	0	36– –>
3	Capella	0.05	45– –>
4	Rigel	0.08	600– –>
5	Procyon	0.34	11.4 END

copy

1	Sirius	−1.5	8.6– –>
2	Canopus	−0.73	98– –>
3	Alpha–Centauri	0.1	4.3 END

deletion

1	Vega	0.04	26– –>
2	Arcturus	0	36– –>
3	Capella	0.05	45– –>
4	Rigel	0.08	600 END

merge

1	Vega	0.04	26– –>
2	Arcturus	0	36– –>
3	Capella	0.05	45– –>
4	Rigel	0.08	600– –>
5	Sirius	−1.5	8.6– –>
6	Canopus	−0.73	98– –>
7	Alpha–Centauri	0.1	4.3 END

sort

1	Alpha–Centauri	0.1	4.3– –>
2	Sirius	−1.5	8.6– –>
3	Vega	0.04	26– –>
4	Arcturus	0	36– –>
5	Capella	0.05	45– –>
6	Canopus	−0.73	98– –>
7	Rigel	0.08	600 END

A1

Answers and Code

Answers are given where we can. Problems of the 'What happens when you run the program on your system?' variety can only be answered by trying it yourself – though in some cases we can offer hints.

1.1 CHAPTER 1

Answer 1.1   Converted into hex it spells **DEADFACE**.

Answer 1.3   604799 is well over the maximum 16-bit integer (32767, §2.2.1 and Table 3.1) but much less than the maximum 32-bit integer. The correct answer is 167 hours, 59 minutes and 59 seconds . If your answer is correct this indicates that your system is using 32-bit integers as the default for **int**: if not then you probably have a 16-bit default. This is quite likely for **C++**. Declaring **total** as **long int** should fix this. Problems are less likely for **Fortran** but if they arise you need to specify a **KIND** – see §3.1.2.

1.2   CHAPTER 2

Answer 2.1   The 1st and 3rd are illegal in **Fortran**, due to length and the initial underscore. **new** and **switch** are **C++** keywords and therefore illegal in that language. **C++** will regard **nucleus** and **Nucleus** as two different variables, but to **Fortran** they will be the same.

Answer 2.2   The techniques used here are discussed in Chapter 10. Considering just the **point** function,

(i) using an extra argument

```
void mypoint(float x , float y , float s)
{
    point(x*s , y*s);
    return;
}
```

(i) using an extra argument

```
SUBROUTINE mypoint(x , y , s)
REAL , INTENT(IN) :: x , y , s
    CALL point(x*s , y*s)
RETURN
END
```

(ii) using global data

```
float s = 1.0;          // global
void set_scale(float ss);
void myplot(float x , float y);
main()
{
    /* ... */
    setscale(2.0);      // or s = 2.0;
    myplot(2.3 , 4.4);
    return;
}
void set_scale(float ss)
{
    s = ss;
    return;
}
void myplot(float x , float y)
{
    plot(x*s , y*s);
    return;
}
```

(ii) using global data

```
MODULE plotting
REAL , PUBLIC :: s
CONTAINS
    SUBROUTINE myplot(x , y)
    REAL , INTENT(IN) :: x , y
        plot(x*s , y*s)
    RETURN
    END SUBROUTINE myplot
END MODULE plotting
```

(iii) using a class

```
class plotter
{
private:
  float s = 1.0;
public:
  void set_scale(float ss)
  {
    s = ss;
    return;
  }
  void myplot(float x , float y)
  {
    plot(x*s , y*s);
    return;
  }
};
main()
{      /* ... */
    plotter pp;
    pp.setscale(2.0);
    pp.myplot(2.2 , 3.3);
      /* ... */
}
```

(iii) using a module with private data

```
MODULE plotting
REAL , PRIVATE :: s
CONTAINS
  SUBROUTINE myplot(x , y)
  REAL , INTENT(IN) :: x , y
    plot(x*s , y*s)
  RETURN
  END SUBROUTINE myplot
  SUBROUTINE set_scale(ss)
  REAL , INTENT(IN) :: ss
    s = ss
  RETURN
  END SUBROUTINE set_scale
END MODULE plotting
```

Answer 2.3   It is far preferable as it uses all the available machine accuracy. There is no direct **Fortran** equivalent because you are not able to use functions in an initialisation.

Answer 2.6   For an integer **i** the following are TRUE if **i** is even

$$((i \% 2) == 0) \qquad \qquad (MOD(i , 2) == 0)$$

Answer 2.8   0, 1, –3, 1, 2, 9, 9, 9, 16

Answer 2.9   0, 4, 3, –3, –6, 4, 3

Answer 2.10   It is parsed as $a = (b = (c = 1.0))$; and thus the first expression evaluated sets **c** to **1.0** and returns the value **1.0**, which is assigned to **b**, etc.

Answer 2.11   0.66..., 0, 6.66..., $2.42 \cdot 10^{24}$

Answer 2.13   (a) = 3 and (d) = 3.0 because of the integer division.

1.3   CHAPTER 3

Answer 3.1 $-89, -28, -1, 364_8, 376_8, 135_8$

Answer 3.2   (a) 0 to 1023   (b) -512 to 511

Answer 3.3   The lower bits add together to give a simple sum. The top bit is given by the sum of the two individual top bits plus any overflow from the lower bits: although this could give a result from 0 to 3, only the bottom bit is kept. If both top bits are 0 (two positive numbers) there can be no overflow for the result to fall in the range of possible values so the final top bit is zero. If both are 1 (two negative numbers) there must be an overflow if the result is to fall in range, so the final top bit (from 1+1+1) is set. If there is one of each (one positive and one negative number) then if and only if the two lower parts are large enough, i.e. if the positive number has larger magnitude than the negative, will the overflow bit add to the original top bit to give a final top bit of zero (from 1+1+0).

Answer 3.6   Using (where appropriate) the procedure for negation of interchanging ones and zeros and then adding 1 gives:
$$101_2 = -4 + 1 = -3, \quad 100_2 = -4, \quad 011 = 3, \quad 100 = 4$$

Answer 3.7   The advantage was that you got a bigger range of floating-point numbers represented for the same number of bits used for the exponent. The disadvantage was that leading zero bits in the significand could be needed, so the precision varied.

Answer 3.9   The assignment sets **k** to zero and returns the value zero, i.e. FALSE, so the output never occurs. The condition **k** $==$ **0** can be TRUE or FALSE.

Answer 3.10   The first is a single character (1 byte), the second a string of length 1 (2 bytes). **'a2'** is illegal.

Answer 3.12   It is quite easy to add a check:

```
if (n > 100)                    IF (n > 100) THEN
{                                   n = 100
  n = 100;                          PRINT * , ' Warning: n set to 100'
  cout << "n set to 100!" << endl;  ENDIF
}
```

With a bit more work you could prompt the user for a correct value (and check that). See §4.12.

Answer 3.14   For **Fortran** you should declare **gas(6,100000)** to keep all 6 elements for each molecule nearby. For **C++** you declare **gas[100000][6]**, likewise.

1.4   CHAPTER 4

Answer 4.1   **(A .OR. B) .AND. .NOT. (A .AND. B).**

Answer 4.2   **!!x_logic** is two operations – it is equivalent to **!(!x_logic)**. But surely NOT(NOT *x_logic*) is just *x_logic*? Yes, but if **x_logic** is any non-zero value (true) then **!x_logic** is 0 and **!!x_logic** is 1. So it maps zero and non-zero to zero and one. The comma operator does this more transparently.

Answer 4.3   **fit_data()** returns a value (1 or 0) which is assigned to **i** and is also the value of the assignment statement and thus determines which of the branches is taken. You could just do this with **if(fit_data())**.... The reason for putting the result in **i** is so that the error handler can use the value to decide which disaster has happened and what action to take.

Answer 4.4   2 and 1 respectively. Show this also with pencil and paper.

Answer 4.5   I walk if it's not raining and I'm on time, or I don't have enough money for the fare.

Answer 4.6   There is little difference. Normally **cin >> i** returns a pointer to **cin**.This is non-zero and evaluates as TRUE, so the **i >= 0** comparison is done. However if the input is unsuccessful, perhaps because the input file is exhausted, **cin >> i** returns the value zero (FALSE): the whole bracket then evaluates as false (without testing **i**) and the while loop stops. So this is actually preferable.

Answer 4.7   A statement can include a type declaration. So you can declare the loop index at the same time as you initialise it, **for (int i=1; i<10; i++)**... Its scope is that of the loop only.

Answer 4.8

```
INTEGER :: k                          REAL FUNCTION f(x)
REAL :: x                             REAL , INTENT(IN) :: x
REAL , EXTERNAL :: F , DF               f = x − cos(x)
x = 0.0                               RETURN
DO k = 1,10                           END FUNCTION f
    x = x − F(x)/DF(x)                REAL FUNCTION df(x)
    IF(ABS(F(x)) <= 0.0001) EXIT      REAL , INTENT(IN) :: x
ENDDO                                   df = 1.0 + sin(x)
   ! ...                              RETURN
END                                   END FUNCTION df
```

Answer 4.9   If **test** is negative then **disc** is undefined in the expression for **x1**.

This is easy to see in the right-hand program: at the **x1** assignment the control must come through the **IF–ELSE** statement, but may come via either branch. In the other the route is less clear.

Answer 4.10   The flag is set to zero if the resource is free. A process wanting to use the resource increments the flag and then checks it. If it's 1 then all is well, and the process uses the resource. Otherwise it decrements the flag, waits, and tries again. When the process has finished it must always decrement the flag to free the resource. If two processes try and book the flag at once, they may lock each other out, but not permanently, if their delay times are different.

1.5 CHAPTER 5

Answer 5.1   The return value is the number of characters successfully printed.

Answer 5.2   **swap2** swaps the values in the function, but they are only copies of the values supplied by the calling program, which keeps the unswapped originals.

Answer 5.3   The first case obviously works. The second might look problematic: if you coded

temp = m;   m = a[m];   a[m] = temp;

that certainly would not work. With, say, **m = 3** and **a[3] = 5** then line 2 sets **m** to 5 and line 3 sets **m[5]** to 3. However the reference is set up *when the call is made*. Changing the value of **i** – strictly, the value in the location pointed to by the reference **i**, which is **m** – does not change the the reference **j**.

Answer 5.4   It should check that its two initial function values have opposite signs. A systematic check across the range given to ensure that there is only one change of sign would be highly desirable. And it should return if it happens to find an exact zero. The range of the search should be printed, and the final value achieved.

Answer 5.5   Replace the **INTEGER :: r , f** by **REAL :: r , f**, and the two occurrences of **1** by **1.0**. A loop would use

```
REAL  FUNCTION  factorial(r)          DO j = 2 , r
    INTEGER :: r , j                      x = x * REAL(j)
    REAL      :: x                     ENDDO
    x = 1.0   ! note that setting      factorial = x
  ! this in the declaration invokes the   RETURN
  ! SAVE attribute – not wanted here   END  FUNCTION  factorial
```

Answer 5.6   Set **x = n** and use one **DO** loop, **j** running from 2 to **min(r , n–r)**, with **x = x * REAL(n–j+1) / REAL(j)**.

1.6   CHAPTER 6

Answer 6.1   The transfer program can discover whether the systems between which it is transporting data use one character for newline (like UNIX) or two characters (like DOS). When it transfers a text file it makes the appropriate substitutions. For a binary file it doesn't. If you get this wrong it only matters if you're transferring between systems of the two different types. A text file transported as binary will have an odd behaviour at the end of every line. A binary file transported as text will occasionally have a spurious byte added or removed.

Answer 6.2   Basically none. They both may signify that following characters are to be treated specially, but in very different contexts.

Answer 6.3   In the **Fortran** language single and double apostrophes are alternative delimiters for character strings. In **C++** **'x'** is a single-byte character, whereas **"x"** is a two-byte string, the first byte being **'x'**, the second being null.

Answer 6.4   Replace the declaration by **char reply[100]**. The input works without being altered. The switch becomes **switch(toupper(reply[0]))**.

Answer 6.5   **strcmp** will compare the five matching characters, and then the **s** of **"Smithson"** with the terminating null of **"Smith"**. These do not match, the difference is −**'s'**, and this is returned. In **Fortran** for, say, **'Smithson' > 'Smith'** the shorter **'Smith'** is treated as having 3 spaces at the end and as **'n'** is (for ASCII) greater than space this is TRUE.

Answer 6.6   If two strings are equal in one character set then even if they are translated into another character set they are still equal.

Answer 6.7   **"edgehog"** and 2 respectively. To get the equivalent of **index** in **C++** you could say:

```
char work[ ]="Hedgehog";
int index = strstr(work , "edgehog") — work;
```

and to get the substring in **Fortran** you can say

```
CHARACTER work*100 , sub*100
work = 'Hedgehog'
i = INDEX(work , 'edgehog')
sub = work(i:)
```

1.7 CHAPTER 7

Answer 7.1   (i) Illegal as the address of an **int** is allocated to a pointer-to-**float** (*ii*) 9,   (*iii*) 46.8,   (*iv*) 8,   (*v*) 9,   (*vi*) 70.2

Answer 7.2   On a typical system: 4 and 20. It will remain unchanged.

Answer 7.4   The **q** => **p** assignment makes **q** point to whatever **p** is pointing to at the time, not forever. So **q** will continue to point to **x** even if **p** is reassigned or deallocated.

1.8 CHAPTER 8

Answer 8.1   Try them and see.

Answer 8.2   It prints **101102103** without any spaces.

Answer 8.3   Try them and see! On ours it gave nonsense.

Answer 8.4   For, say a 6 by 5 matrix

WRITE(6, '(I5 , 5F9.3)') ((j , array(j , i), i=1,6) , j=1,5)

Answer 8.5   Because s is an array, and arrays are passed through their addresses automatically. For ordinary variables you have to instruct the compiler explicitly to take the address.

Answer 8.7   It could contain something like

```
char s[100];
for(int i=0;  i<100 && ((s[i]=getchar())  !=  '\n');  i++);
s[i] = NULL;
```

Answer 8.9   These implementations take a negative number to indicate the end of the data:

```
ofstream out;
out.open("data.store");
int i;
while(cin >> i , i > 0) out << i;
out.close();
```

```
OPEN(45 , 'data.store')
DO
   READ * , i
   IF(i < 0 ) EXIT
   WRITE(45 , '(I8)') i
ENDDO
CLOSE(45)
```

Answer 8.10   For the global-data approach you place the definition **FILE* fff;** or **ofstream fff;** or **INTEGER fff** into a header file to be included (in **C++**) or a module to be used (in **Fortran**). For the object approach you define an object with **fff** as private data, and write necessary functions for it.

Answer 8.11

```
FILE* f;                                 ifstream f;
f = open("data.016" , "r");              f.open("data.016" , ios::nocreate);
```

Answer 8.12

```
#include <fstream.h>
int pages;
char name[20] , title[100];
int recsize =  sizeof(pages)  +  sizeof(name)  +  sizeof(title);
// note that the operator sizeof does not require brackets :
//        eg     sizeof pages  +  sizeof name  +  sizeof title;
ifstream f;
f.open("results" , ios::binary);
f.seekg(99*recsize);
f >>  setw(4)  >>  pages;
f >>  setw(20)  >>  name;
f >>  setw(100)  >>  title;
    /* ... */
```

1.9   CHAPTER 9

Answer 9.1   Remember that default numbers in **C++** are doubles while those in **Fortran** are single precision. Unless you cast to floats the **C++** results are likely to be about 10^{-17} rather than about 10^{-8}, but still not equal.

Answer 9.3   You will probably find that the (real) value of **epstest** evaluates as zero, thanks to the extra precision used in intermediate arithmetic.

Answer 9.4   The test works for an **epsilon** that is a factor of two smaller than it was with the + sign. This is because of the number representation. For a base-10 analogy: you can accurately represent $(1 - 0.001)$ with three digits, but for $(1 + 0.001)$ you need four.

Answer 9.5   Here is one way of doing it; with doubles the exponent range is ± 308 and hence no underflows are possible.

```
double a0 = 5.29E−10, epsilon0 = 8.854E−12, echarge= 1.602e−19;
double energy = −pow(echarge/a0 , 2) / 8.0 / M_PI / epsilon0;
```

But $e^2/(8\pi) = 1.02 \cdot 10^{-39}$ which is out of range for a float/single-precision variable and could underflow to zero depending how it is programmed.

Answer 9.6

```
PROGRAM water_mean
INTEGER , PARAMETER :: (N=4)
REAL :: c(1:N) , cmean , fmean
CHARACTER (LEN=10) :: fmt
DATA c/3.782 , 5.540 , 5.769 , 7.066/
    fmean = SUM(c) / REAL(N)    ! first use the false mean 12680.000
    fmt   =   '(A , 5F10.4)'
    PRINT fmt , ' c vector   = ', c
    PRINT fmt , ' false mean c    = ', fmean
    PRINT fmt , ' residuals = ', c − fmean
    c = c + 12680.0             ! now recover the original data
    cmean = SUM(c) / REAL(N)
    PRINT fmt , ' c vector   = ', c
    PRINT fmt , ' raw mean c    = ', cmean
    PRINT fmt , ' residuals = ', c − cmean
    STOP ' water mean ended'
END PROGRAM water_mean
```

Answer 9.7   Try it: the first method is susceptible to accuracy problems as β^2 gets close to 1.000.

```
PROGRAM muon_momentum
REAL :: muon_mass = 105.0    ! in units of GeV/c**2
REAL :: p(6) = (/1., 20., 50., 100., 200., 500./)
REAL :: beta , E1 , E2
DO k = 1,6
    beta = p(k) / sqrt(muon_mass**2 + p(k)**2)
    E1 = muon_mass / sqrt(1.0 − beta**2)
    E2 = sqrt(p(k)**2 + muon_mass**2)
    PRINT '(5ES15.5)' , p(k) , beta , E1 , E2
ENDDO
STOP
END PROGRAM muon_momentum
```

Answer 9.8

```
//      this code uses references to do addressing, see section 5.2.3
#include <iostream.h>
void meanvar(float &mean , float &var , float data[ ] , int n);

main()
{
    float data[ ] = { 3.782 , 5.540 , 5.769 , 7.066 };
    float mean , var;
    meanvar(mean , var , data , 4);
    cout << "mean value = " << mean << "variance =  " << var << endl;
    return 0;
}
void meanvar(float &mean , float &var , float data[ ] , int n)
  {       // n data values in the array
    mean = 0.0;
      for(int i=0;  i<n;  i++)
      {
          float delta = (data[i] − mean) / float(i+1);
          mean += delta;
          var += float(i)*delta*delta − var/float(i+1);
      }
    return;
  }
```

Answer 9.9

```
PROGRAM  Horner_polynomial          REAL  FUNCTION  Horner(p , N , x)
REAL , EXTERNAL :: Horner           ! N is the highest polynomial power of x
REAL :: v , g , a(0:3)              ! array p holds the coefficients
DATA  a/ 2.000 , 4.142 , 3.492 , 6.670 /    INTEGER :: k , N
  v = 1.0                          REAL     :: p(0:N) , x , horn
  g = 1.0 / Horner(a , 3 , v)        horn = 0.0
  PRINT *, v, g, a                   DO k = N, 0 , −1
  v = 2.0                             horn = p(k) + horn * x
  g = 1.0 / Horner(a , 3 , v)        ENDDO
  PRINT *, v, g                      Horner = horn
END  PROGRAM  Horner_polynomial     END  FUNCTION  Horner
```

1.10 CHAPTER 10

Answer 10.1   The data for a histogram should contain: the number of bins, their widths, the number of entries in each bin, the number of entries that fell outside the plot (overflows and underflows), perhaps a title for the histogram and a title for the axis.

What can you do with a histogram? You can define it, you can add a data value to it, and you can plot it. You can extract the values in any bin, or range of bins. You can find the mean and standard deviation (and other properties) of the data. You can fit functions to it, compare it to other histograms, and there are probably several more things to do that you can think of.

Answer 10.2

```
vector vector::rotate(double theta)
   {
       vector v;
       v.x = x*cos(theta) − y*sin(theta);      v.y = x*sin(theta) − y*cos(theta);
       return v;
   }
```

Answer 10.3   The class will be similar to the one already given, with three data values **x** , **y** , **z** instead of two. The vector product function would be

```
vector vector::vector_product(vector a)
   {
       vector v;
       v.x = y*a.z − z*a.y;     v.y = z*a.x − x*a.z;     v.z = x*a.y − y*a.x;
       return v;
   }
```

Answer 10.4   Integers are straightforward and are in the language anyway. For complex numbers you have to add the two parts (real and imaginary) separately. For matrices you have to add each element, which means you have to know how many there are. For strings you probably want '+' to denote concatenation of two strings.

Answer 10.5   You need to define

```
vector vector::operator+(vector);
vector vector::operator−(vector);
vector vector::operator*(double);          // scale vector by factor
double vector::operator*(vector);          // scalar product
```

Answer 10.6   You can just add the line **for(int i=0; i<n; i++) data[i] = 0.0;** to do the zeroing. Or **data[i] = given;** for initialisation to a given value.

You can implement this by defining two constructors, **vector::vector(int n);** for initialising to zero and **vector::vector(int n , float given);** for initialising to a value. These overloaded functions would be called by **vector a(3);** and **vector b(3 , 1.0);** respectively. It is shorter and more compact and altogether better to use a default argument, and define **vector::vector(int n , float given = 0.0);**.

Answer 10.7   Because **data** is a pointer, and we want to delete the object pointed to, not the pointer itself.

Answer 10.8   With the program of this section, the class definition needs the extra line

friend double sprod(vec u, vec v);   double sprod(vec v);

The functions are

```
double sprod(vec u , vec v)              double sprod(vec v)
  {                                        {
      return u.x*v.x + u.y*v.y + u.z*v.z;     return u.x*v.x + u.y*v.y + u.z*v.z;
  }                                        }
```

and they would be called by statements like

double ff = sprod(a , b);   double ff = a.sprod(b);

Answer 10.10   Replace **struct** by **class**. Then add **public:** within each definition, at the start. And you also have to replace **class circle:shape** by **class circle: public shape** etc.

The member-selection operator **.** is left-associative so the first thing evaluated is **arrow.scale(2.0)**. This presumably doubles the vector size as desired, and returns the contents of the pointer to the object itself, here **arrow**. So the evaluation continues with the expression **arrow.rotate(45.0).shift(−10. , −10.)**. And they are thus executed in sequence.

The ASCII Character Set

The ASCII character set is given in the following table:

	0	1	2	3	4	5	6	7	8	9	A	B	C	D	E	F
0	nul	soh	stx	etx	eot	enq	ack	bel	bs	ht	lf	vt	ff	cr	so	si
1	dle	dc1	dc2	dc3	dc4	nak	syn	etb	can	em	sub	esc	fs	gs	rs	us
2		!	\	#	$	%	&	'	(	)	*	+	,	−	.	/
3	0	1	2	3	4	5	6	7	8	9	:	;	<	=	>	?
4	@	A	B	C	D	E	F	G	H	I	J	K	L	M	N	O
5	P	Q	R	S	T	U	V	W	X	Y	Z	[	\	]	^	_
6	`	a	b	c	d	e	f	g	h	i	j	k	l	m	n	o
7	p	q	r	s	t	u	v	w	x	y	z	{			}		del

The first hexadecimal digit is given by the row, the second by the column. Thus $7A_{16}$ is the lower-case character **z**. Characters below the blank-space character 20_{16}, decimal 32, are non-printing. Their two- or three-letter acronyms sometimes stand for meaningful words such as 'escape', 'null', 'bell', 'backspace', 'horizontal tab' 'vertical tab' but many are lost in history. They can – sometimes – be entered into a keyboard by pressing the 'Control' key at the same time as a character with an ASCII code 40_{16} higher. Thus Control-H gives **bs** – 'backspace'.

The character $7F_{16}$ was taken as the deleted character because when you had a paper tape with a wrong character on it you could remove that character by punching all of the extra holes.

A3

Bibliography

Adams J C, Brainerd W S, Martin J T, Smith B T and Wagener J L (1992) *FORTRAN 90 HANDBOOK, Complete ANSI/ISO Reference*, McGraw-Hill, ISBN 0-07-000406-4

Barlow R J (1989) *Statistics, A Guide to the Use of Statistical Methods in the Physical Sciences*, John Wiley and Sons Ltd., ISBN 0-471-92295-1

Cary J R, Shasharina S G, Cummings J C, Reynders J V W and Hinker P J (1997) Comparison of C++ and Fortran 90 for object-oriented scientific programming, *Computer Physics Communications* Vol 105 pp 20–36

Dijkstra E W (1968) Goto statement considered harmful, *Communications of the ACM* Vol 11, pp 147–148

Hatton L (1995) *Safer C:: Developing Software for High-integrity and Safety-critical Systems*, McGraw-Hill, ISBN 0-07-707640-0

Kernighan B W and Ritchie D M (1988) *The C Programming Language, ANSI C*, 2nd ed. Prentice-Hall, ISBN 0-13-110370-9

Kerrigan J F (1994) *Migrating to Fortran 90*, O'Reilly & Associates, Inc., ISBN 1-56592-049-X

Knuth D E (1968) *The Art of Computer Programming* Volume 1. Addison-Wesley, ISBN 0-201-03809-9

Metcalf M and Reid J (1992) *Fortran 90 Explained*, Oxford University Press, ISBN 0-19-853772-7

Metcalf M and Reid J (1996a) *Fortran 90/95 Explained*, Oxford University Press, ISBN 0-19-851888-9

Metcalf M and Reid J (1996b) *The F Programming Language*, Oxford University Press, ISBN 0-19-850026-2

Press W H, Teukolsky S A, Vettering W T and Flannery B P (1993a) *Numerical Recipes in C*, 2nd edn. Cambridge University Press, ISBN 0-521-43108-5

Press W H, Teukolsky S A, Vettering W T and Flannery B P (1993b) *Numerical Recipes in Fortran 77*, Cambridge University Press, ISBN 0-521-43064-X

Press W H, Teukolsky S A, Vettering W T and Flannery B P (1996) *Numerical Recipes in Fortran 90*, Cambridge University Press, ISBN 0-521-57439-0

Rice J R (1983) *Numerical Methods, Software, and Analysis*, McGraw-Hill, ISBN 0-07-066507-9

Stroustrup B (1993) *The C++ Programming Language*, 2nd ed. Addison-Wesley, ISBN 0-201-53992-6

Stroustrup B (1997) *The C++ Programming Language*, 3rd ed. Addison-Wesley ISBN 0-201-88954-4

Index